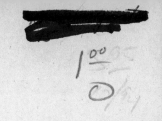

HarperPerennial

A Division of HarperCollins*Publishers*

THE POWER
OF PLACE

.

HOW OUR
SURROUNDINGS
SHAPE OUR
THOUGHTS,
EMOTIONS, AND
ACTIONS

.

WINIFRED GALLAGHER

A hardcover edition of this book was originally published in 1993 by Poseidon Press, an imprint of Simon & Schuster, Inc. It is here reprinted by arrangement with Simon & Schuster, Inc.

First HarperPerennial edition published 1994.

Designed by Barbara M. Bachman

Library of Congress Cataloging-in-Publication Data

Gallagher, Winifred.
    The power of place : how our surroundings shape our thoughts, emotions, and actions / Winifred Gallagher.—1st HarperPerennial ed.
        p. cm.
    Originally published: New York : Poseidon Press, c1993.
    Includes bibliographical references and index.
    ISBN 0-06-097602-0 (pbk.)
    1. Environmental psychology. 2. Man—Influence of environment.
    I. Title.
    [BF353.G355 1994]
    155.9'1—dc20   93-41798

94 95 96 97 98 CW 10 9 8 7 6 5 4 3 2 1

■

TO MIKE,

WHO TAKES CARE

OF THE

OUTSIDE

■

ACKNOWLEDGMENTS

For their erudition, time, and good nature, I am deeply grateful to the many scientists who shared their research with me and reviewed my understanding of it. For their particular insight and patience, I am especially indebted to Drs. Kerry Feldman, Myron Hofer, Stephen Kaplan, Peter Suedfeld, Ralph Taylor, and Thomas Wehr.

CONTENTS

PART III ■ SYNCHRONY

. . . he saw the flaming, restless nest of the fire a short distance away, and the silhouettes of people around it, and someone's hand adding a branch. The crickets leapt crepitating; from time to time there came a sweet whiff of burning juniper; and above the black alpestrine steppe, above the silken sea, the enormous, all-engulfing sky, dove grey with stars, made one's head spin, and suddenly Martin again experienced a feeling he had known on more than one occasion as a child: an unbearable intensification of all his senses, a magical and demanding impulse, the presence of something for which it was alone worth living.

—Vladimir Nabokov, *Glory*

INTRODUCTION

.

THE

SCIENCE

OF PLACE

L AST SPRING, I spent several days sealed off from the sweet palmy
swelter of New Orleans in a series of frigid polyester conference
rooms, listening to men in white coats discuss the latest developments
in brain science. Weary of sci-fi scanning techniques and neurotrans-
mitter balances, I treated myself to a lecture by Mihaly Csikszentmihalyi, a professor of psychology at the University of Chicago best
known for his improbable best-seller, *Flow;* despite its easygoing title,
this rather difficult, scholarly book methodically explores the parameters of what the author terms "optimum experience" and the rest
of us call a good time. As I had hoped, Csikszentmihalyi was not
concerned with the workings of pills or the measurement of rats, and
his introductory remarks, which centered on his earliest forays into
his chosen area of inquiry, went straight for the jugular of behavioral
science. "As a small child, I wondered why most of the otherwise
knowledgeable, accomplished adults who surrounded me seemed to
have almost no idea about how to live a satisfying life," he said. "It
was clear to me even then that the answer wasn't money or power
but, somehow, the ability to control and enjoy one's experience."

A bit of unaccustomed reflection on one dimension of our experience suggests that the answer to that perennial child's question—If grown-ups know so much, why aren't they happy?—is increasingly bound up with the places in which we spend our lives. Many of the eclectic researches that support this commonsensical idea are less discoveries than rediscoveries of principles that our forebears considered obvious. Throughout history, people of all cultures have assumed that environment influences behavior. Now modern science is confirming that our actions, thoughts, and feelings are indeed shaped not just by our genes and neurochemistry, history and relationships, but also by our surroundings.

More than two thousand years ago, Hippocrates' observation that our well-being is affected by our settings was established as a cornerstone of Western medicine. The healers of antiquity had no idea that the malaria parasite is carried by mosquitoes, but they noted that the residents of hill towns were healthier than those from marshy regions, and concluded that the problem was the "bad air"—*mal aria*—of such places. Of all the environmental influences on a person's state, however, "it is chiefly the changes of seasons which produce diseases, and in the seasons the great changes from cold or heat," wrote the father of medicine, adding that "Such diseases as increase in the winter ought to cease in the summer, and such as increase in the summer ought to cease in the winter. . . ."

The centuries of literature on the relationship between mood and the seasons comprise a striking testimony to science's venerable association of behavior and environment. One of the most fundamental and enduring principles of classical medicine positively encouraged analogies between our internal and external climates. Physicians believed that the action of the four humors, or body fluids, determined everything from a person's constitution to his character. Because the balances of yellow bile, black bile, phlegm, and blood, which corresponded to the four elements of fire, earth, water, and air, were also related to summer, fall, winter, and spring, an individual's physiological and behavioral changes were inevitably viewed in the context of the sun's. In the second century A.D., Aretaeus prescribed that "Lethargics are to be laid in the light and exposed to the rays of the sun, for the disease is gloom"; in the fourth century, Posidonius observed: "Melancholy occurs in autumn, whereas mania in summer."

Classical science's propensity for viewing a person's state in its environmental context persisted down through the ages. In the seventeenth century, the English scholar Robert Burton, who suffered from bipolar disorder, in which profound depressions alternate with chaotic bursts of mania, compiled his exhaustive *Anatomy of Melancholy*. This text includes some stereotypical assumptions about climate and national as well as individual temperament that remain commonplace: "Hot countries are most troubled with . . . great numbers of madmen. . . . They are ordinarily so choleric in their speeches, that scarce two words pass without railing or chiding in common talk, and often quarreling in their streets. . . . Cold air in the other extreme is almost as bad as hot. . . . In those northern countries, the people are therefore generally dull, heavy, and [include] many witches, which [some] ascribe to melancholy." Two hundred years later, this tendency to see connections between behavior and its setting still prevailed among the first practitioners of the infant science of psychiatry.

Around the turn of the twentieth century, the wisdom of the ages concerning the relationship between place and state was eclipsed by technological and cultural changes so rapid and vast that social scientists still debate our ability to adjust to them. In one of the least remarked of these transformations, the Industrial Revolution drew the West indoors. Turning away from the natural world, huge populations gravitated toward a very different one made up of homes and workplaces that were warm and illuminated regardless of season or time of the day—although even on a rainy morning, it is brighter outside than inside with the lights on.

Society quickly adapted to its new indoor urban environment. Only a hundred years ago, the overwhelming majority of Americans lived in the country, while today, most cluster in metropolitan areas. Like other living things, however, our species has evolved over millions of years to respond to the cycles of the earth and sun with predictable biochemical and behavioral changes. Environmentally minded scientists have begun to question the trade-offs we unwittingly make in order to live sealed up inside an artificially heated, cooled, and lighted world that is structured around economic rather than biologic concerns. They point out, for example, that in the West, exposure to the sun's bright light has become erratic in duration and timing for the first time in history, and they suspect that the fact that most of us are no longer wakened by the dawn, drawn outdoors for much of the day by our way of life, and lulled to sleep by darkness helps explain why up to a third of us suffer from sleep or mood problems, or both. What is startling about that observation is not the idea that light has to do with mental and physical health, which was accepted a hundred years ago, but the fact that science forgot about it.

That the importance of light was obscured even from scientists has partly to do with a profound change in the modern Western attitude toward time itself. Over the millennia, our species developed a cyclical sense of time, still marked by recurring solstice festivals, such as Christmas and Hanukkah, which bring light into the dark winter, and the spring celebrations, such as Easter and Passover, which celebrate the sun's return. The gradual replacement of this mode of time perception by the more recent historical, linear concept helped conceal seasonal mood shifts from modern medicine.

Just as society began to measure time by its own doings rather than nature's, individuals increasingly looked inward rather than outward for insights into their behavior. This propensity was encouraged by two milestones in twentieth-century science. First came Freudian theory, which emphasized the overwhelming importance of a person's internal psychological processes, heavily shaped by the past, in determining his ways. Metaphorically, the inward orientation of psychoanalysis, rooted in the thinking of its early Eastern European forefathers, reflects something of the enclosed, restricted environment of the *shtetl*, whose residents could not always move about freely. In such a culture, what psychoanalysts would later disparage as a "flight into health" was not necessarily an option. Convinced of the therapeutic primacy of insight and inner change, Freudians were skeptical of the idea that altering one's milieu, say, where one lived, might also have merit. That kind of thing was, they said, "running away from your problems," even though the people who ran away sometimes felt better.

As the psychoanalytic influence broadened, promoting soul searching as the royal road to wisdom and well-being, the rapid growth of psychopharmacology gave rise to another hypothesis, one in which all depended on inheriting the right neurochemistry. Both psychoanalysis and drugs, which are often seen as antithetical, emphasize the individual's internal processes as the determinants of mental health, and pay little attention to the external environment, as earlier schools of thought had.

By promoting a false dichotomy between the influences of biology and environment—often narrowly interpreted as meaning only the social setting—academe has also helped obscure the synchrony between behavior and its milieu. However, the cutting edge of scientific inquiry is tearing away at conceptual barriers between actions and the contexts in which they occur, even at the simplest levels of life. The study of molecular genetics, for example, reveals that what a cell will be is determined not just by what is in it but also by who its neighbors are; through various constituents it is sensitive to, the gene's microenvironment influences its workings. "Every cell in your body has the same genes, and yet there are incredible differences in their shapes and functions," says Myron Hofer, director of the department of developmental psychobiology at the New York State Psychiatric Institute and the author of *The Roots of Human Behavior*. "Why is that? Because the 'developmental plan' everybody talks about very much depends on where cells happen to find themselves. And conservative nature, which uses the same biological processes over and over throughout development and evolution, has not discarded the actions of molecules and cells in producing the behavior of organisms."

Even the simplest microorganism depends on environmental interactions to survive. Hofer points to the migrations of the tiny typhoid bacillus: in order to live, it must swim to a place rich in the nutrient it wants, stop, and remain there, finding its way by the reactions of receptors on its cell membrane to chemicals that send it into different states. Because we tend to describe behavior in psychological terms, says Hofer, "we say the bacillus 'knows' enough to avoid the wrong and be attracted by the right places, and even has a 'memory' of about a tenth of a second to enable it to make such 'judgments.' " When it comes to our own behavior, this tendency to psychologize makes it especially hard to see the biological and environmental influences on it. "We may prefer to view our behavior in psychological terms, but its origins are biological, both in the individual and in the evolution of the species," says Hofer. "The biology of behavior concerns the four elements of molecule, cell, organ, and organism, and the physical environment is important from the simplest level up through any stage in our development."

Like those of other living things, our structure, development, and behavior rise from a genetic foundation sunk in an environmental context. Yet while we readily accept that a healthy seed can't grow into a plant without the right soil, light, and water, and that a feral dog won't behave like a pet, we resist recognizing the importance of environment in our own lives. Nonetheless, says Hofer, "place, which is a good term in that there's not a lot of baggage to it, is important to human beings right from the beginning. Sperm and eggs are very clearly influenced by their immediate surroundings. When changes in the mucus of the uterine environment facilitate the process, sperm have a better chance of fertilizing the egg. In some way, the mother even decides which sperm gets to her ovum. During wartime, for example, X-bearing sperm are somehow favored. And if an egg is not in the right spot, it can't form a placenta. The same principles apply when a zygote is only a few cells as when a whole baby is interacting with its world. Nature and nurture cooperate in producing behavior, and the organism has a lot of say about what it's going to get out of its environment."

A good bit of that "say," however, depends on an organism's being in the right place at the right time. If it is not, despite normal genes, its physiological and behavioral development can be skewed. Siamese cats raised in a cold environment develop brown fur while those reared in a hot one are white, for example, and chicks incubated in an environment in which the eggs aren't rotated end up with malformed leg joints. Similarly, kittens raised in the dark stumble, fall, and collide when they are put in the light, and rats reared in dull environments have smaller brains than those raised in complex ones. "When you're in that straitjacket of thinking in dichotomies of nature or nurture, you're going to make mistakes," says Hofer. "We can talk about them as components of a system, but to see genes and environment in opposition—no. I wouldn't even say they interact, exactly. The proper analogy is closer to what happened when the Pilgrims arrived in America. They didn't simply duplicate England, although its culture exerted lots of influence. The United States is really a process that developed from what the settlers were predisposed to do and the new environment's enhancement of and constraints on that, from the climate to the presence of the Indians. It's impossible to say that one or the other determined the nation."

Hofer's enthusiasm for the subject of environment and behavior has developed over years of iconoclastic research on our most important relationship; although the mother-infant bond has traditionally been considered a social one, his studies show that it is woven from the sensory strands of its physical, flesh-and-blood milieu. In laboratories and neonatal intensive care units, along power lines and in addiction treatment centers, in tropical forests and polar outposts, other researchers from disparate disciplines are also uncovering hidden dimensions of the role our surroundings play in our well-being. An Olympian business, science requires gazing fearlessly into the dark, gaping, fire-breathing maw of the nature of things and, in a facts-just-the-facts voice, reporting back only what can be measured, weighed, or counted. Although such objectivity and restraint are practical and elegant, scientists' formal discourse usually leaves out some of the juicy parts of their thinking. An untidy interdisciplinary subject such as the effects of places on behavior permits them to relax and indulge in some speculation, and to combine perspectives with those of peers from other fields.

These days, any big conference of scientists concerned with the future of our planet or species includes presentations and discussions of aspects of the relationship between people and places. At a recent annual meeting of the American Psychological Association, for example, two papers suggested something of the complexities to be revealed by increasingly sophisticated investigations of this feedback system. One study was an analysis of the ways in which lively and dull interiors affect mood and performance. When the subjects' responses to a stimulating, plant-filled, homey setting and a grim, institutional one were contrasted, the only reaction they all shared was a decline in vigorous activity and increased feelings of fatigue in the austere environment. This finding suggests that one reason staying in the hospital furthers recuperation is that it makes patients feel tired, which means it is easier for them to stay in bed and rest; they perk up when they get home because the environment there is so much more stimulating. More significant, however, the subjects' other responses to their surroundings varied according to certain of their personal traits, such as the tendency to "lose" oneself in a task or to be easily distracted. In other words, an everyday setting that inclines one individual to feel and function well can push another in the opposite direction. The next step in this research is the development of an "environmental sensitivity scale" or questionnaire that will systematically measure individual reactions to a range of ambient conditions. Equipped with good data on which, or how many, people are sensitive to lighting, spatial arrangements, noise, and other ordinary features of our surroundings, architects, office managers, doctors, and the rest of us will be better able to create more supportive, personalized environments.

Marcel Proust experienced such extreme reactions to the environment that he spent much of his time sequestered in a cork-lined boudoir. Aware that the immune and nervous systems are particularly responsive to environmental stimuli, scientists who study the links between psychological and physical health and the environment aren't surprised that the hypersensitive introvert also suffered from allergies and hay fever. A second paper presented at the APA meeting noted that of the 10 percent of the population vulnerable to depressive illness (a tendency that, along with creativity, often runs in certain families), the majority also have a history of allergy. Among subjects from this group, physical and behavioral responses to the environment can be particularly tangled. For example, those inclined to get depressed in the winter also generally feel blue in humid atmospheres with high pollen counts; linked fluctuations in the levels of beta-endorphins, associated with mood, and histamines, involved in allergic reactions, may be responsible for their warm-weather distress. Furthermore, investigation revealed a strong correlation between allergy, particularly hay fever, and the personality trait of shyness, defined by scientists simply as a negative reaction to novelty. This research suggests not only that there may be a distinct subgroup of shy people who are vulnerable to specific allergies and affective disorders, but also that when such a person enters a dusty, unfamiliar room, he's likelier than others to have an allergic reaction. Although a seemingly small thing such as being in a familiar or unfamiliar setting can make a big difference in certain of our individual responses, few of us, including research scientists, as yet appreciate the importance of environmental influences on our states.

These two studies quietly demonstrate that like our choices about relationships or careers, our decisions about where to live or work can have a significant if often unsuspected impact on our well-being, whether through subtle means, such as lighting and plants, or more directly, through agents such as allergens or pollutants. Think for a moment about our chemical environment—the air we breathe, the water we drink. Over the past twenty years, it has been drastically altered by technology, but we don't understand the complex ways that the resulting changes could affect us. It is possible that chemical stressors are pushing certain people over the threshold for developing particular physical or psychological problems. In the future, a trip to the doctor may well involve an evaluation of such environmental components of our health.

Just as the world around us affects our behavior, our thoughts, emotions, and actions affect our surroundings. When asked to predict the most important environmental influence on behavior in the twenty-first century, researchers almost invariably give the same answer: urbanization, or making places citylike without necessarily making cities. Yet it would be equally accurate to say that urbanization is also the most important behavioral influence on the environment. The technological and social changes associated with this unprecedented worldwide development mean that before we superficially adjust to a new, lower status quo, our ever-adaptable species must understand what a good environment really is, in a community as well as a forest, in an office and school as well as a home. Burdened with increasingly complex social roles, we need places that support rather than fragment our lives, places that balance the hard, standardized, and cost-efficient with the natural, personal, and healthful. To secure this kind of environmental quality in a rapidly changing world, we must put the principles emerging from the multidisciplinary science of places into practice on local and global levels.

In its hard-edged way, the research of environmentally minded scientists often confirms our own softer perceptions and intuitions, both individual and cultural, concerning the relationship between people and places. My own absorption in the subject began with the purchase of a small wooded property in upstate New York. From my first glimpse of the one-room schoolhouse commanding its sunlit site with all the authority of a small Greek temple, I knew I was home. Practitioners of the eponymous Chinese form of geomancy would attribute this feeling to the place's *feng shui*, or "wind and water." Not easy to nail down in Western terms, *feng shui* more or less corresponds to what we call ambience, or a place's distinctive atmosphere. Resting on a gentle shelf above a dirt road, the building is protected from bad weather by the low mountains behind it, while the rill and brook that flow in front attract great blue herons, trout, deer, wildflowers, and other harbingers of benign *chi*, a ubiquitous energy that permeates places as well as people and other living things. According to Chinese tradition, the proper balance of *chi* ensures good *feng shui*. For whatever reason, this country place immediately struck me as just the thing to counter some of the environmental stress—bad *feng shui*, if you will—that life in a big city entails. My Manhattan street corner is a good example. Several times a day, the neighbors recoil from the shriek of tires, and even the crunch of metal, as two cars meet on a collision course at this intersection. The traffic isn't unusually heavy there, but two wide boulevards cross hard by an avenue coming in at a diagonal, resulting in three streaming rivers of undisciplined *chi* that converge just where the cars so often do.

Behavioral scientists don't employ *feng shui* terminology, but their method of analyzing environments in terms of the stimulation they afford boils down to much the same thing. Whether we think about *chi* or sensory input, we all seek a comfortable level of arousal from our settings, one that is neither so low as to court boredom nor so high as to invite anxiety. Over the course of a day, a week, and a year, most of us seek places that provide different degrees of stimulation. Arriving at the schoolhouse on Friday, I collapse and wait for the rhythms of nature to work their magic on my fried urban nerves; after a few days of staring at the creek, I'm eager for the Broadway boogie-woogie again. Like weekenders from the time of ancient Rome, I have learned to regulate psychology with geography.

After discovering the benefits of running away from my problems, I began to exploit the therapeutic potential of my everyday surroundings at times when a flight into health wasn't feasible. Whether the strategy is to clean the house from one end to the other or walk a few miles of city streets, I've found that exposure to larger systems of physical organization, preferably while working the body, almost invariably brings glimmerings of hope and purpose. As William James observed while writing of another source for these two sustaining sensibilities, religion involves "the belief that there is an unseen order, and that our supreme good lies in harmoniously adjusting ourselves thereto."

Along with sharpening the awareness of how "a change of scene," as we often put it, makes one feel like "a new person," country life aroused my interest in some peculiarities of the unseen order that must have intrigued Mr. James. Just after I bought my hundred-year-old house, the plumbing backed up. The local expert surveyed the dismal situation in the bathroom, then inquired where the septic tank was located. I had no idea. He asked if I wanted him to find it, and despite gloomy thoughts of the bill for having a backhoe destroy what passes for my lawn, I nodded, willing to pay anything to right the terrible wrong indoors. Rather than going back to town for the heavy equipment, however, the plumber repaired to the woods, returning with a Y-shaped branch. Holding it by the arms, he strolled about the property until the stick's tail dipped sharply downward, as if an invisible hand were pulling it to the earth. Beneath a flat rock lay the tank. As he set about his work, the plumber said he had picked up the knack of dowsing from his father and other veterans in the trade who were understandably cautious about where they started digging in this rocky soil. "On these old properties, lots of new owners can't find their cesspools, drain lines, even their wells," he said. "Time after time I've picked up a forked stick—just about any one will do—and headed right to the spot. I can't tell you why that's so, but it is."

Scientists willing to speculate about why it is so think that like many animals, including birds, dolphins, and whales, certain people might have the physiological capacity to detect very tiny variations, frequently associated with underground ore or water, in the geomagnetic field. As if from a giant magnet, this energy emanates from the earth's core and wraps around it like a blanket; irregularities in its warp and weft account for some very peculiar events. One night in the country, not long after the dowsing episode, I tried unsuccessfully to dismiss thoughts of extraterrestrials while watching balls of light glide about some three feet off the ground in a boggy area near the creek. Some digging in scientific journals suggested the glowing orbs were probably "ghost lights"—luminous displays of natural electricity likeliest to occur in marshy spots or areas riddled with geomagnetic irregularities.

If any one experience inclined me to spend several years exploring the power of place, it was a trip through a part of America especially rich in geophysical peculiarities, where bumper stickers read: "Feel the Magic." The Four Corners area of the Southwest, of which Santa Fe is the unofficial capital, has a very old reputation among the Hopi, Navaho, Pueblo, and Spanish colonial cultures as a "sacred place" possessing special psychoactive properties. For hundreds of years, one of its top tourist attractions has remained the tiny village of Chimayo, the "Lourdes of America" that draws thousands of visitors each year during Holy Week. Long before Spanish missionaries dedicated a charming adobe chapel to the Madonna there, nearby Native Americans had believed that the springs and surrounding soil had special healing powers. This theory got some impressive support in the early nineteenth century, when a Christian visitor to the shrine saw a burst of light shoot from a spot on the ground. Since the auspicious fireworks, piles of crutches, eyeglasses, and braces of those claiming to have been cured by the Virgin's "holy earth," exposed by a hole in the chapel's floor and free for the taking, have accumulated in the *santuario*.

At 3 A.M. on the morning after visiting Chimayo, I set off on a pilgrimage to a different kind of sacred place: the Jimez hot springs, carved into a mountain ridge in the high country above Los Alamos. My host on the freezing, moonless hike up to the site was an esteemed member of the Santa Clara Pueblo. Trusting entirely in what certainly seemed like his supernatural power to lead on through the pitch dark, up slippery slopes and across invisible logs spanning roaring streams, I finally glimpsed a tier of falls dropping into three natural basins in the obsidian cliffs. Plunging through steam clouds rising in the frigid air, we sank gratefully into the 100-degree water. At 4 A.M., our only neighbors were some southern bikers and their girlfriends, who sported and mated, smoked pot and sipped Jack Daniel's in one of the pools below. Fearing that their raunchy behavior offended my companion, whose people consider this spot holy, I offered a prissy apology. Amused, he shook his head. "This place likes what they do," he said. "However, we usually come here to cleanse ourselves.

Heal ourselves. As we tell our young people, sometimes you can't just watch the video. There are special things in the places all around us, but you may have to work hard to see them."

Bobbing gently above the vast valley as rose began to tint the sky behind the purply Sangre de Cristo range, I found that all the special things to see and otherwise sense fostered a special state of mind. Hardly for the first time, it struck me that we are only infinitesimal flickers in the great scheme of things, but the emotion this inspired was altogether different from feeling like just another gnat during rush hour. This powerful perception of being a very small fish in a very big pond, called the "diminutive effect" by environmental psychologists, has been deliberately cultivated by the architects of Gothic cathedrals and Nazi stadiums alike. My sense of total immersion in the surroundings was further augmented by the wall of sound created by the thundering falls, and perhaps by certain unusual properties of a setting situated at a heady altitude in an area riddled with geophysical peculiarities. Certainly the sulfurous fumes exuded by the springs would have been familiar to the Delphic oracle, said to have derived a good bit of her inspiration from whiffing volcanic gases. In short, some alchemy of natural grandeur framed in an unfamiliar Native American context, a few objective somethings-or-other, and a riot of neurochemicals set off by sleep deprivation, cold air, novelty, hot water, and some voyeurism produced an experience I'm happy enough to attribute to my presence in a magical place. Fortunately, speculating about some of its parts doesn't detract from the mystery of the whole.

The reluctance to dismiss heightened experiences in special places as entirely subjective is shared by James Lovelock, the British scientist who developed the "Gaia hypothesis." This influential concept, which forms the core of my oldest daughter's high school biology course this year, approaches the earth and its processes as a unified living organism rather than as a grab bag of separate biological and geophysical systems. By profession, Lovelock is obliged to remain in the realm of the rational, but he doesn't discount the unusual experiences he has enjoyed in certain natural settings merely because science can't yet explain them. One such place is Brentor, an ancient volcanic site near his home.

Lovelock has written that when he climbs this strange hump commanding a vista of the Atlantic, the Bodmin moor, and the Dartmoor cliffs, he sometimes experiences "a sense of presence. Not extrasensory, but something perceived by the senses that can neither be seen, heard, or felt in the usual way. It would be easy to attribute to this sensation the recognition of something sacred. A momentary contact with some entity larger and greater than the mind." Attempting to explain the derivation of this sensation, he cites a physicist colleague's special interpretation: the senses convey to the brain far more information than we can consciously be aware of; it is the totality of all that undifferentiated input that we perceive in a general way as ambience. At special places such as Brentor and the Jimez springs, this diffuse essence is somehow stronger and more poignant.

Soon after returning from New Mexico, I started to do the research for this book. Every once in a while, someone who knew what I was up to would say something along these lines: All right then, where's the *best* place? There is no way to answer that question without asking "For whom?" The only universal truth I've discovered during the past few years' work is that the recipe for the good life that Mihaly Csikszentmihalyi and all the rest of us imagined as children calls for being in the right place at the right time as often as we can manage.

OUTSIDE IN

.

PART I

.

1

.

A DAY
AND
NIGHT WORLD

No MATTER where you are in Alaska, or when, winter is never far away. Most visitors come during the brief season of 78-degree "heat waves" and nearly nonstop sunlight. Yet hikers, campers, and anglers can set out on a radiant summer morning only to spend the afternoon shivering in a dark gray, freezing rain that is just the mildest premonitory lashing of winter's tail. Even sedentary tourists content with observing the Albert Bierstadt fantasy of Mount Denali framed by a windshield sense something of what December is like from Alaskan settlements. Towns and villages have the no-frills, battened-down air usually associated with military bases, and the insides of buildings evoke portholed spacecraft, designed to seal out an alien environment. A place of violent extremes—the longest and shortest days, the harshest winters and most dazzling summers—Alaska is a giant living laboratory for the study of light and temperature, the two greatest environmental influences on living things.

"Especially at first, the winter here is an eerie experience," says John Booker. As a psychologist, a medical sociologist, and associate dean of health sciences at the University of Alaska in Anchorage, he has a keen professional interest in how Alaskans respond to sea-sonal extremes. "The cities stick to the same schedule people keep in the Lower Forty-eight. If you work indoors, that means you go inside at eight A.M. when it's pitch black, and when you come out at four or five P.M., it's pitch black again. You never see the day. Up north in Fairbanks, they get less than four hours of daylight at the solstice. Above the Arctic Circle, say, in Barrow, the sun sets around Thanksgiving and doesn't rise until mid-February. Here in Anchor-age, the sun comes up at ten-thirty and sets at three-thirty. For maybe an hour in the middle, there's some fairly bright sunlight, but it has no warmth. You can't feel the sun on your skin at all. It's cold, too—thirty below is about average. By mid-December, what you really have is a nice sunrise immediately followed by a sunset that lingers for three and a half hours. Only at the end of February do things start to move back toward a day-and-night world."

The first scientist to record how people react to Alaskan-style seasonal extremes was Frederick Cook, a nineteenth-century Amer-ican naval officer, doctor, and polar explorer. At one point, his ship was trapped in ice, and the sailors saw no sun for sixty-eight days. Describing what would be known a hundred years later as seasonal affective disorder (SAD), Cook wrote that the men "gradually be-came affected, body and soul, with languor. . . . The root cause of these disasters was the lack of the sun." When the sailors were exposed to artificial light, he noted, their good spirits returned. The long, brilliant days of polar summer, however, could be almost too salubrious. As soon as the months-long winter night ended, Cook reported, the Eskimos were seized by a spring-feverish euphoria during which they abandoned themselves almost entirely to courtship.

Cook's research is far more than a mere account of exotic ways in bizarre places. According to Thomas Wehr, a research psychiatrist who is the chief of psychobiology at the National Institute of Mental Health and its leading authority on environmental influences on behavior, "his observation that human reactions to environmental changes in the extreme realms of the poles are merely amplifications of the subtler seasonal changes we all experience in more temperate regions is a major legacy of the great Victorian scientific expeditions that included Darwin's. Pretty much anything you can say about mood sickness can be traced back to normal responses. It's a matter of degree. There are people whose energy and motivation are simply lower in winter and higher in summer, and others who get really depressed or manic at those times."

The origins of the influences of light on our activity are rooted far back in the evolutionary past. The survival of our species has depended on matching the workings of our bodies and minds to the demands of day and night. Most animals can't function equally well in darkness and light, and must concentrate their efforts either on night, as rodents do, or day, as we do. "It's almost as if our planet has two worlds," says Wehr. "Depending on whether we're inside at night or outside during the day, we have to change our natures and become different kinds of animals. The daytime creatures who must venture out into the field are colder and brighter, aggressive and seeking. At night, when we conserve our energy, we stay in our burrowlike homes, warm and insulated from outside stimuli. Just as our workday environments—offices, factories, farms—are quite different from the domestic one, so are our ways in those two places."

This gift for adjusting our states to our places has been aeons in the making. Like all life on earth, Homo sapiens evolved in the warm glow of the sun. Because it changes through the course of the day and the year in such a predictable way, sunlight is an ideal stimulus for the synchronization of our biological rhythms. "Let's say you're a northern animal that has to rely on temperature to tell you what season it is so you know when to mate, hibernate, sprout horns, change colors, and give birth," says Wehr. "Suddenly, there are a few mild days in February. You'd be in real trouble if you decided spring had arrived, because overnight it could drop to ten degrees again."

Throughout the millennia, daily and seasonal fluctuations in light and temperature have caused us to develop the biological means to anticipate them. These responses of the body to its environment, which ebb and flow in approximately twenty-four-hour cycles known as circadian rhythms, incline us to be active and alert in the brightness of day and to rest in the darkness of night. Our daily physiological and behavioral shifts are intimately connected to our seasonal ones because the brain, equipped with a light meter that gauges the day's illumination and a biological "clock" that measures the day's length, uses information about light conditions to determine the time of the year. These internal timing mechanisms are so important to our functioning that they tick away even in the womb.

No one knows exactly how the brain translates light into patterns of behavior. As the most popular scenario has it, light rays strike the retina, triggering nerve impulses that travel to the brain along pathways that function strictly as light meters and have nothing to do with vision in the sense of image processing. These neural impulses are delivered to the front of the hypothalamus, which Wehr calls "the Grand Central Station for all kinds of important behavior, including sleep, mood, hunger, and libido." The propinquity of behavioral matériel in the close quarters of the hypothalamus, where, for example, the body's main thermostat sits next to the entry point for neural light-meter fibers, helps explain some of the common origins and overlap of problems seemingly as different as insomnia, depression, jet lag, and menstrual troubles.

From the busy, crowded hypothalamus, light information is transferred to the pineal gland, once known as the mystical "third eye" that interprets the outside world to the interior one. Scientists are finding that this still somewhat mysterious brain structure is indeed extraordinarily sensitive to environmental cues, and probably mediates or reinforces human circadian rhythms. In response to certain light conditions, the pineal produces a hormone called melatonin, thought to induce the sensation of sleepiness. To stay in synchrony with the external world—resting when it is dark and keeping active when it is bright—we need a high level of melatonin at night and a low one in the morning. Ideally, the hormone's tide begins to rise around 9:30 P.M., peaks after 11, and drops off in the middle of the night so we can wake up early. A few moments' exposure to candlelight will shut down production of melatonin in a rat, but it takes a much brighter light of about 300 to 2,500 lux to turn it off in a person; a typical sunrise, which measures about 750 lux, is more than enough to do the job. Research on animals has shown that melatonin levels also vary with the seasons, rising in winter and declining in spring in response to the changing conditions of light.

The disturbances in circadian cycles that make us yawn by day and fidget by night can result from an inadequate amount of light exposure, as may happen in winter, or an adequate dose at the "wrong" time of day, as occurs when we cross time zones quickly. Along with these external stimuli, the stubbornly individualistic inclinations of our biological clocks—the neural time-keeping mechanisms that schedule our circadian rhythms—can help disrupt things from the inside as well. Like other timepieces, these must be set to synchronize with the sun. The body clock of a person bereft of light cues, say, deep in an underground cave, will start ticking to its own rhythm; soon the circadian cycles will follow, running perhaps on a twenty-three- or twenty-five-hour pattern. The inner clock of the "night owl," who wakes up groggy each morning and gets friskier toward evening, wants to run long—perhaps on a twenty-five-hour cycle. To remain synchronized with the sun's schedule instead, the clock requires a correction mechanism: light. Exposure to bright light in the morning will subtract an hour from the night owl's internal schedule by programming his melatonin level to drop off before it would do so on its own. On the other hand, the internal cycle of the "lark" runs shorter than the sun's, so he ends up nodding off right after dinner and waking at 5 A.M. Light exposure in the evening could *add* time to his internal day.

When Michael Terman, a psychologist who is the director of the light therapy unit of the New York State Psychiatric Institute, wants to be sure of a good night's sleep, he turns on his "tropical dawn machine," an experimental computerized light device he invented to recreate Caribbean-style dawns and sunsets in bedrooms around the globe. The electrical sun starts to sink as he gets into bed at 11 P.M., and by 11:25, he is fast asleep. Sunrise occurs promptly at 7 A.M., when he wakes feeling alert and refreshed. "I spent a winter with the machine and I was always asleep and awake at the right times, despite my chaotic habits," he says. "The artificial twilight at night is like a hypnotic agent. Halfway through the sunset, one is calm and ready to sleep. Light can be a photic sleeping pill—a healthful one."

The connection between a spell of poor sleep, fatigue, and melancholy is so strong that depression can be defined as a long bout of exhaustion and the emotions it inspires: helplessness and hopelessness. Most people who have flown more than a time zone or two from home have had a small taste of what struggling with that condition is like. In the variation on depression known as jet lag, certain things that happen in the traveler's body every day at certain times— the release of particular neurochemicals, for example, or the rise and fall of temperature—continue to occur on the schedule set by light changes back home. For the few days it takes the brain to start operating on the new light signals, the traveler is lost somewhere between Topeka and Tokyo, glassy-eyed and woolly-brained during working hours, jazzed up at bedtime. Given a few weeks' duration, these clashes between what the body wants to do and what the world demands of it would turn into the complex psychological and physical state that, no matter what its cause, is summed up as depression.

Of all the behavioral gripes Alaskans have in winter, the commonest is feeling tired—the physical side of depression. "That body sensation subsequently turns into an emotional experience," says John Booker. "If every winter you have trouble going to sleep and then wake up at four A.M., even though you haven't had a good night's rest in a week, you're going to be a lot more irritable and have more social problems. And when you're tired and grouchy, you cut down on fraternizing and going out, which burrows you even further into your bad mood."

You don't have to live in Alaska to endure—and inflict on others— a winter-long bad mood. For large numbers of Americans living north of the fortieth or fiftieth parallels—roughly from Washington and Oregon on one coast to the New England states on the other—the short dark days between November and April add up to a long nightmare. Their annual immersion in a black pit of melancholy and inertia is aggravated by difficulties with appetite, sleep, libido, and cognition: their minds don't work right, and neither do their bodies. Statistics on how many people are affected by SAD vary from study to study, but in 1988, a collaborative survey conducted in four different latitudes within the United States, each with its different length of day and angles of winter sun, showed that while only 2 percent of Florida residents had severe seasonal mood problems, the figure jumped to 6 percent in Maryland and New York, and 10 percent in New Hampshire.

Assuming the worst, for a long time researchers did not even bother to find out how many Alaskans are troubled by winter. Finally, in 1989, Booker and Carla Hellekson, a psychiatrist then affiliated with the University of Alaska at Fairbanks, decided to find out exactly what toll their long, dark winters exacted on their fellow citizens' well-being. They found the data they needed in the results of a comprehensive health survey conducted earlier in the Fairbanks area. During home visits lasting an hour and a half, interviewers had thoroughly grilled 310 subjects, representative of the community in socioeconomic status, sex, and race—black, Native Alaskan, and white—about every aspect of their health. When Booker and Hellekson analyzed the information on the impact of winter and summer on the subjects' well-being, they got some surprises.

The first was a pleasant one. "Considering the seasonal conditions in Fairbanks, it was a bit of a shock to find that just under nine percent of our subjects get seriously depressed in the winter," says Booker. "That figure is similar to New England's." Although comparatively few subjects were almost disabled by winter, the study revealed a far more widespread and overlooked problem: practically everyone experienced some degree of seasonal distress. While 19 percent of the Fairbanks group classed as "subsyndromal" were bothered by some but not all serious SAD symptoms, a whopping 50 percent underwent spells of low energy, overeating, and poor sleep.

Alaskans are hardly the only Americans who struggle, often unsuspectingly, with degrees of semi-SAD. After conducting the New York area segment of the four-latitude study, Michael Terman found that 18 percent of his subjects fit into the subsyndromal category, and another 26 percent were troubled by some sort of February doldrums. "When you add these two groups to the people who become seriously depressed," he says, "you see that half the population of the New York area—more than four million people—undergoes some kind of behavioral change in winter."

During the almost endless days of Alaskan summer, figuring out what to do when is almost as difficult as, if far more pleasant than, the same quandary in winter. "A researcher who wanted to assess drinking problems among Native people went to an Inupiat town and asked standard alcohol-abuse questions, like 'Do you ever drink before noon?'" says Kerry Feldman, an anthropologist at the University of Alaska at Anchorage. "What does that mean in a place like Barrow, where there's no noon at all for three months and almost round-the-clock daytime in summer?" Because the nonstop light show can make falling asleep and keeping calm in July as hard as waking up and staying active in December, some Alaskans and tourists resort to wearing sunglasses by day and blocking their bedroom windows with aluminum foil at night.

The connection between superabundant light and mania is not nearly as well established as the link between inadequate light and depression. Mania, the high end on the continuum of mood, is generally less well understood than the low, in part because far more people suffer from melancholic than from manic symptoms. (The basic depressive illness is manic depression, or bipolar disorder; some victims experience both the highs and lows, most only the lows, and a few only the highs.) Nonetheless, says Wehr, "we know light is an antidepressant and that antidepressant drugs can trigger mania. We also know that hospitalization for mania goes way up in the summer, that some winter depressives get manic then, and that light therapy can cause mild mania. It's a good bet that mania has to do with light."

Evidence of high spirits, sometimes too high, in the Fairbanks study's subjects during spring and summer supports the idea that seasonal mood changes tend to recur in the two seemingly different yet related forms of summer mania and winter melancholy. A high percentage of those who get depressed in winter not only say that summer is the best time of the year for them, but also act differently then. For example, other Alaskans stick to a stable pattern of alcohol use throughout the year; the SAD-prone, however, usually drink much less than average, only approaching the level of the general population in the depths of winter and at the peak of summer. "The people most affected by the seasons are not those prone to drink, period," says Booker. "For them, it's a sign of acute environmental stress. It seems that individuals who are particularly sensitive to light conditions may very well overshoot normal adaptation both in the winter and the summer."

One particular group of people—musicians, writers, painters, sculptors, and their relations—is especially prone to both the highs linked to abundant light and the lows caused by too little. Studies that have looked for connections between creativity and mental illness show that artists are particularly susceptible to such mood swings. "Goethe wrote 'It is a pity that just the excellent personalities'—he included himself—'suffer most from the adverse effects of the atmosphere,' " says Norman Rosenthal, who is the director of the NIMH's seasonal studies program. "Many of the creative do experience the seasons with great intensity, and are often inspired by these changes to do great works." Although inspiration may come in the summer and winter, the works themselves are likelier to be produced when conditions are less extreme; for example, Handel and Mahler tended to have highs in summer and lows in winter but were most productive in spring and fall. In the life of Vincent Van Gogh, Wehr sees a record of a search for an environmental cure for such recurring mood swings. "Van Gogh probably suffered from SAD, among other problems," he says. "His work shows his obsession with light and color. Some paintings are really black and ominous, and others are full of sun and stars. Even after he migrated from the cloudy gray Belgian mining town where he was born to much brighter Provence, however, he seems to have gotten manic in the summer and low in the winter."

Keeping track of his own patients' mood swings was the thing that first interested Wehr in studying environmental influences on behavior. Aware that a third of the people treated for psychiatric illness have regularly occurring ups and downs, he noticed that one of his NIMH patients got depressed every 35.5 days—not 35, not 36. "Then, like clockwork, she'd get undepressed," he says. "She was unaware of this precise timing, and had lots of psychological explanations about why she felt better or worse. Meanwhile, I knew that it had been thirty-five and a half days since she felt well. That kind of gave me the creeps. It made me wonder how much we fool ourselves about why we feel what we feel and think what we think. I began to suspect that our philosophies and that kind of thing are affected by biological and environmental factors that we still don't fully understand."

Wehr's suspicions were confirmed in 1980, when Herb Kern, an engineer who suffered from recurrent melancholia, showed up at the NIMH. Kern, accustomed to thinking analytically, had kept careful records of his mood cycles and theorized that his problem had to do with the seasons. Alfred Lewy, then a research psychiatrist at the institute, was already investigating the effects of light on mood, and his findings intrigued Wehr. When Kern and his data arrived, he says, "we really kind of got going with that information." With NIMH's powerful imprimatur, the long-neglected study of the effects of light on behavior took a giant step forward. Its most obvious application is the increasingly familiar phototherapy panel that helps SAD victims dispel winter's gloom: a few days of the right amount of brightness at the right time resets errant biological clocks, and repeated exposure keeps them running smoothly. Some researchers think that light may someday treat depressions that don't appear to follow a seasonal pattern.

Until antidepressant drugs were developed in the 1960s, melancholia was thought to have a psychological etiology rooted in guilt feelings. Since then, it has usually been at least partly ascribed to an imbalance of one or more neurotransmitters—serotonin, norepinephrine, dopamine, and 200 to 300 other juices that carry the electrochemical signals that regulate behavior. Of the legion of transmitters, scientists can identify about 50, and really know something about only a half-dozen or so. "It's no wonder that the same patient can be diagnosed by three different clinicians as having three different disorders and still be prescribed the same serotonergic agent!" says Terman. "One reason transmitter research leaves something to be desired is its tight focus on the internal chemistry of the sufferer, which ignores other, external influences on behavior." While it is just as simplistic to say that light is involved with all depression as to put the whole blame on transmitters, for more than a decade, Daniel Kripke, a psychiatrist who's the light expert at the University of California at San Diego, has effectively used phototherapy to treat some depressives who don't suffer from SAD. His good results have at least, as he puts it, "raised the question of how distinct SAD is from other forms of depression."

No matter what the specifics of their etiologies, behavioral problems that respond to phototherapy will be easier to treat in the future. The basic light regimen prescribed at the New York State Psychiatric Institute for people with serious SAD is thirty minutes' exposure to a panel that dispenses 10,000 lux, which is considerably more convenient than the more traditional approach, which uses 2,500 lux and takes two hours. Soon phototherapy won't even require a "light box"; NIMH researchers are already experimenting with a visor that could deliver medicinal rays to the wearer on planes and in hotels and offices. Michael Terman looks forward to the day when this kind of high technology will allow an astronaut to benefit from an earthlike lighting cycle while in space, and a New Yorker to avoid jet lag by adjusting his circadian rhythms to London time before leaving home. The same type of light treatment can be a great boon to shift workers and others plagued by sleep problems caused by erratic schedules. Daniel Kripke, who directs a sleep-disorders clinic, cautions that while using light to reset wayward biological clocks could certainly be a better therapy for insomnia than drugs, it is not a universal panacea for a problem that can have many different causes. "Light helps some poor sleepers, particularly those who don't get outdoors very much," he says. "For example, elderly women often stay inside because they're afraid of crime or injury, and a lot of the marked difficulty they have with sleep may be caused by light deprivation. Getting outside more, particularly toward evening, can be very beneficial for such people."

The most avant-garde area of research on light's behavioral effects concerns reproduction. Because other animals must bear young when their survival is likeliest, their mating behavior is seasonal, cued by changing conditions of sunlight. Although humans are the only creatures who can mate when they please, Frederick Cook noted that Eskimo women virtually stopped menstruating during the dark winter; a century later, his observation is being tested by an ongoing study of Native Alaskan women in northern villages, conducted by the University of Alaska at Anchorage. In a lab setting, scientists on Kripke's team have also showed that modest illumination can dramatically affect menstruation: a 100-watt bulb used as a night light corrects cycles that are overly long. It may be that even the pale light of the moon can influence female rhythms.

The ancient notion that lunar changes affect the ebbs and flows of human behavior, from luck to sexuality and even sanity, remains so pervasive that half the college students polled in one survey said they believed that people act strangely during a full moon. However, says Rosenthal, "despite the long history associating the moon's cycles with 'lunacy,' we still have no scientific evidence for it. If they do exist, the moon's behavioral effects might be due to light or to gravitational effects on our body fluids—reasonable ideas, but unproved."

One of the foundations of the belief in a link between women's rhythms and the moon's is the fact that the average menstrual cycle is the same as the lunar one—29.5 days. It has been hypothesized that long ago, women menstruated at the full moon, and that their cycles are so various today because of mankind's interference with the subtleties of natural light. "We now know that low levels of light at night can adjust long cycles," says Terman. "It's possible that in the future, such treatment could be refined by simulating the moonlight cycle in the bedrooms of women with irregular menstrual cycles."

In the future, light therapy may also prove to be a special boon to women affected by what most people call premenstrual syndrome, or PMS, and scientists insist on calling late luteal phase dysphoric disorder. True PMS is not just some bitchiness and bloating a few days before a period, but a virtually disabling funk lasting at least a week. Barbara Parry, a research psychiatrist who is a colleague of Daniel Kripke's at the University of California at San Diego Medical Center, measured the melatonin levels of some women troubled by PMS, and found they had lower balances throughout their cycles, as well as daily levels that dropped off too soon; when treated with two hours of evening light, they scored significantly lower on tests that measure depression than they had before. Kripke speculates that one of women's most agonizing problems may in some cases also eventually be treated with light. "If light affects menstruation, it may be involved with infertility, too," he says. "Should it turn out that light regularizes cycles by preventing ovulation, it could also turn out to be an extremely useful method of contraception, and make the rhythm method more efficient. The study of light's effects on reproductive endocrinology is really just beginning."

Along with these practical applications of light science, the widespread use of an environmental agent to change behavior has profound philosophical implications. Most patients who end up in SAD clinics have already had lots of traditional psychotherapy that did nothing to relieve their misery. The fact that 80 percent respond to treatment with light presages a new order of drug-free, talk-free psychiatry. Even the manner of the researchers who study environmental influences on behavior differs from that of their more traditional colleagues. It is hard to imagine whining about neurotic troubles to Wehr. "If you exposed yourself to light, and three or four days later you felt very much better, your whole attitude changed, your functioning improved dramatically, and you became more creative, what's to talk about?" he says. "Some patients need antidepressants in addition to light to feel well, and they all need to discuss how mood swings can insidiously affect their lives. But after a few initial sessions, psychotherapy is like skiing—people learn best with a tune-up lesson now and then. Unless something's cooking, I see my average seasonal patient about four times a year."

Unearthing the environmental roots of what feels like an emotional problem may turn out to be the biggest hurdle facing many people in need of help. William Mills, a physician practicing in Anchorage who pioneered the modern treatment of hypothermia and cold injuries, says that as the holidays approach, "a lot of people up here say, 'We hate Christmas. We are screwed up because Christmas is coming with all that fuss.' Of course, Christmas just happens to be in the middle of the darkness. At least now there's some awareness that those who have a hard time in the winter aren't psychiatric basket cases." John Booker's experience with drug-abuse prevention programs also inclines him to think that seasonal mood problems should be viewed less as an individual hang-up and more as a public health issue. "It's not like there's just a certain group out there with a bad gene who gets a mental illness called SAD," he says. "Human response to light lies on a continuum, and some people are simply more sensitive to the changes that affect everybody." Terman agrees, favoring a national effort to educate people about environmental influences on behavior. "All of us show seasonal peaks and troughs in mood," he says. "This response to the external world is a general human tendency—the norm, not the exception. Because substantial numbers have an exaggerated reaction, seasonal behavioral changes should be a public health concern, especially when we can treat them so simply and successfully with light."

2

.

THE CLIMATE
INDOORS

"Y ou look around Anchorage in July, and you could be a lot of places," says John Booker. The window of his office at the university duly frames the very picture of the idyllic American campus, complete with green lawns, flower beds, and even a sunlit march of mountains. "If you saw how much trouble it is to maintain the population here a few months from now, however, you would understand that this is a fairly artificial environment, and the farther north you go, the more that's true. Maintaining a business-as-usual, nine-to-five attitude here in December puts us at odds with what's going on outside to an extreme degree, as well as requiring a lot of money and effort. To keep half a million people in Alaska year round is something like operating in an outpost on the moon."

For part of each year all Alaskans—Native, white, and black— are deprived of light, but not everyone gets upset. "I've never heard an Eskimo complain about the winter weather or its effects on mood, even in Barrow, which is just godawful," says Irvin Rothrock, a psychiatrist in private practice in Fairbanks, and just about everyone who lives up North would probably agree. "It could be an evolutionary thing. Over the centuries, those subject to seasonal mood illness just didn't survive, which eventually eliminated it from their gene pool." The hypothesis that over time, an extreme environment could influence its inhabitants' genes not only makes good sense, but also seems to be supported by a survey of eighty Siberian Natives very similar to American Eskimos in ancestry. The then Soviet researchers who studied them found no sign of winter depression, in contrast to a high rate among eighty workers transplanted from other parts of the country.

There's no doubt that certain people are inherently more vulnerable to depression, seasonal or otherwise. Booker has collaborated with the ex-Soviets on studies of health issues raised by extreme environments; because hundreds of thousands of their workers must be relocated to Siberia, where much industry is concentrated, establishing criteria for those best able to adapt to the cold, dark winters is important. One of the most reliable gauges turns out to be a candidate's sleep patterns. "Scientists there are much more physiologically minded than we are when it comes to evaluating behavior," he says. "They believe that someone who gets up early, can sleep at any time, and copes well with schedule changes is likelier to live longer and stay free of mood disorders and ailments linked to stress, such as heart disease and ulcers, than a night owl who doesn't feel well in the morning, is bothered by changes in routine, and suffers from jet lag."

Remarkable advances in the field of genetics have accustomed us to the notion that we inherit certain biological influences on our behavior. While this once-controversial concept is now scientifically respectable, the idea that living at odds with the natural world—courtesy of electric light, air-conditioning, and central heating—might have consequences strikes us as mere romanticism. Yet when they analyzed how the various peoples of Fairbanks cope with the stress of winter, John Booker and Carla Hellekson made a discovery that challenges traditional assumptions about the nature of our adaptation to our environments and about why people from different places are different: their statistics showed that proportionately as many urban Eskimos suffered from seasonal mood swings as did their black and white neighbors. "Unlike the Natives in the rural villages here and the Siberians in the Soviet study, who live in the traditional outdoor manner, the Fairbanks Natives live in apartment houses and work nine-to-five jobs," says Booker. "Even for Eskimos, there's a link between the individual's way of life and his adaptation to the seasons."

In order to bloom, the biological seeds of winter depression must be sowed in the right environment. Activated by light deprivation, the disorder is fed by conflicts between the victim's internal clock and his society's alarm clock: he wants to sleep when the nine-to-five world is awake and bustle about while it slumbers. As the Fairbanks data imply, this cultural dimension means that the more a society lives at odds with what is happening in nature, the more cases of SAD it will have. Urban Alaskans pride themselves on their ability to stick to the conventional American nine-to-five schedule and carry on no matter what it is like outside. The citizens of Fairbanks, who refer to the somewhat more temperate coastal metropolis as Los Anchorage, are particularly renowned for their doughtiness. "During the coldest three or four weeks up here, the Puritan ethic really goes insane," says Jerry Mohatt, a psychologist and dean of the College of Rural Alaska at the state university's Fairbanks branch. "By December, you hear a lot of 'God! When is February going to come?' The institutions we've imposed require a lot of energy, time, and daylight, and even though it's dark and sixty below, we try to proceed as if everything were okay. Parts are snapping off cars because it's so cold, but we bus the kids to school and try to operate businesses. It would be much wiser to shut down and have a vacation period, but people up here like to think that we can handle anything. That attitude is a real contrast to the way Native peoples deal with what's going on outside, because they consider the environment to be a legitimate part of their behavior and psychology."

Other trained observers confirm this impression. "In the rural villages, the Natives subject themselves to much less duress in the winter than we do in the cities," says Rothrock. "Their lives are sure not tied to the clock, or even to the day. In winter, an Eskimo might skip *this* morning and get up at seven o'clock *tomorrow*." Booker agrees that life in the villages is much more in sync with the realities of winter. "Nothing starts at eight in the morning," he says. "Around ten, maybe eleven, things stir for a few hours, and then tail off until evening. People may not sleep eighteen hours straight, but they sure slow down. They make use of the day for what it is, and that's it."

Just because heredity can't explain the different behavioral reactions to the environment shown by different groups doesn't mean a general *human* tendency to slow down in winter hasn't been perpetuated through genes because of some potential usefulness. "When the sun goes down right after it comes up, there's a tendency to shorten your hours of activity," says Rothrock. "The first winter or two after I came here to Alaska from Kansas, I was ready to turn in by afternoon. Once the sun had been down a couple of hours, it was bedtime." According to one adaptive rationale of this reaction, the real challenge winter poses isn't shorter days, but less food and fuel. Faced with diminished resources, it's advantageous for many animals to conserve energy by sleeping more, living off stored fat. SAD research suggests that Homo sapiens is among them; fatigue was the leading winter complaint cited in the Fairbanks study, and while everyone gains weight in winter—about five pounds is average—the depressed often put on twice as much. "Like other animals, lots of people sleep and eat more every winter and have less energy then," says NIMH seasonal researcher Norman Rosenthal. "The lowered mood associated with the change from summer to winter light could be a carryover from the survival mechanism of hibernation. We too might have semihibernated, back in our cave days. More women are affected by winter depression, so it's also logical to speculate that a period of inactivity during a time of environmental stress would be adaptive for pregnancy and childcare. In general, one can see that in the Far North, it could be a very good thing to be inactive and to overeat and oversleep in winter."

Rosenthal's adaptive hypothesis certainly jibes with the seasonal cycles of the Native Alaskan way of life. "The people will tell you that in the old days, when their elders lived very traditionally, it was important to expend as little energy as possible in winter," says Mohatt. "They didn't want to increase their food and fuel needs beyond their resources, which could vary from hunting season to hunting season. Conservation was particularly crucial because the community had to take care of the families of less successful hunters, which is sometimes still the case." In the Native villages, winter's adaptive lethargy is followed by an equally useful burst of activity in spring and summer, when the increased intensity and duration of light reverse the neurochemical changes evoked by December drear. In Fairbanks at the solstice, the sun barely dips below the horizon, creating brilliant days that taper into an almost endless twilight. In the rural villages, people work eighteen or twenty hours at a stretch, catching and preserving fish and game for the winter to come.

The Eskimos' sensible attitude about schedules and the seasons is one mark of a culture that has evolved in great intimacy with its surroundings. Their animistic religion helps make a harsh world comprehensible and even beautiful, solidifies the group, and encourages conservation of resources, particularly in times of scarcity. Adaptation to the howling desolation of Arctic winter begins early, when children are trained to stay alert to their surroundings and to gauge and avoid risks. Dressed in heavy gear invented many centuries before down sleeping bags and insulated boots, they are as warm in their snug microenvironments as people in a temperate climate wearing light clothes. Indoors as well, their culture helps them counter the potential boredom and irritation of the long winter's night with festivals to lighten mood and supply a sense of time. To mimimize friction in close quarters, mishaps are greeted with laughter rather than anger.

Perhaps most important, Native Alaskans see winter as a time to kick back and have some fun, the oldest and best antidepressant. "Once winter arrives, the Natives spend a lot of time visiting and enjoying what we'd call small talk," says Mohatt. "It's really the art of conversation—they're great talkers and jokers." Alaskans who aren't part of such a congenial culture find being cooped up indoors far more stressful. "When people talk about 'cabin fever,' they usually mean an upset caused by isolation and not being able to get out of the house," says Rothrock. "An old-timer who spent most of his life in the bush told me that one winter, he just couldn't stand being in his cabin alone anymore, so despite really terrible conditions, he hiked over to see his nearest neighbor some miles away. When he finally got there and looked in the window, the other man was talking and gesturing and laughing to himself, so he just turned around and went home again. I'm not sure how often that kind of situation comes up these days, but a lot of men don't work in winter because they can't find jobs. Everyone ends up in a small house, getting on each other's nerves. If there's also a good deal of drinking to pass the time, things can snowball into a real bad scene, and they do, fairly frequently."

If seasonal shifts in energy are a general human tendency, the wise would not only relax the business-as-usual schedule in winter but also adopt another Eskimo habit. Regardless of temperature, a rural subsistence culture requires people to go outside for certain periods, during which they get the advantage of whatever light there is. "My hunch is that people who have an active life-style are less disturbed by winter, no matter what their ethnicity," says Booker. "The Alaskans who think it's the best part of the year are always the ones out running dogs or skiing." Rothrock agrees. "The people who do well here tend to be active types who are outdoors all year round—they like to hunt, ski, and fish. At night, they get involved in things at home. We have more bookstores here in Fairbanks than in a town three times as big in the Lower Forty-eight."

Lacking a team of huskies or even decent snow, many other Americans find little reason to linger outdoors in inclement weather, which cuts back on our already scanty light exposure and further lowers spirits. Research makes clear that come December, in Fairbanks, Boston, or Paris, it's a good idea to get out and soak up some daylight, even if the prospect sounds gruesome. Emphasizing the importance of getting outside in the daylight no matter what, Daniel Kripke observes that a colleague in Basel effectively treated some winter depressives by insisting that no matter how bad the Swiss weather was, they had to take an hour's walk. "The waiting rooms at SAD clinics empty out if we get a few anomalous springlike days in February, probably because people attracted outdoors by the balmy air end up getting more light, which activates them and fights depression," says Thomas Wehr. "The temperature is really organizing their behavior, which is an important concept in seasonal behavioral changes. When you're depressed, you further reduce exposure to already deficient winter light by spending more time sleeping and lying around indoors."

Having conducted anthropological studies in the tropics as well as the Arctic, Kerry Feldman is utterly unsurprised by the contrast in the incidence of winter malaise caused by the different ways of life among the state's urban and rural Natives. "Environment never *determines* behavior and has only rarely had significant effects on physiology or heredity, mostly because culture rushes in to inter-vene," he says. "Here's a timely example. When government re-searchers came to Alaska to measure the increase in the rate of skin cancer linked to the polar hole in the ozone layer, they were surprised that the figure was considerably lower than anticipated. In calculating their predictions, they had failed to consider the fact that Alaskans wear more clothes than people from temperate climates, which pro-tects their skin. The environment can impose certain limits on what you do, but how you react to them is up to you and your society. Two different groups can live in very similar environments yet have very different behavior, because the effects of their surroundings are filtered through their culture."

Even in the most salubrious climate, an indoor life-style can mean light deprivation. Using a special photosensitive computer his sub-jects wear on the wrist for several days at a time, Daniel Kripke has piled up data on several hundred Californians' light exposure and activity levels on a minute-by-minute basis. He has found that young people, such as the students on his UCSD campus, are outdoors most, but even they only get an hour or two of direct sunlight a day. In contrast, office-bound adults get less than a half hour. With more leisure, healthy elderly men log an hour and a half a day, but older women get only a half hour. "Light exposure varies with a person's location, age, sex, and occupation," says Kripke. "But even in sunny San Diego, it's surprising how little light people get. If they aren't outdoors very much here, it's less in other places."

That even the darkest of those other places have been settled is testimony to the fact that unlike other creatures, people don't have to make do with natural settings. The sun is not our only source of heat and light, and our ability to warm and brighten whatever place we find ourselves in has enabled us to prosper even in environments like Alaska. As Paul MacLean, the wizard emeritus of NIMH and pioneer of evolutionary psychobiology, points out, "If human beings arose in Africa, their excursions north could only have been made possible by learning to use fire, which allowed them to keep warm, prolong the day, and protect themselves from wild animals."

The interior worlds we create, from log cabins to airplane cabins, have climates that affect our well-being as surely as the climates outside. This often-overlooked principle was a foundation of a splinter group of the Quaker movement known as the Shakers, whose name derives from their spiritual practice of ecstatic dancing. Although otherworldly in religion, the Shakers were pragmatic and forward-thinking, and their zeal for progress led to many beautiful as well as useful inventions, including the flat broom, the clothespin, and the circular saw. Their most prescient legacy, however, might be the illumination of interior environments, which they felt could give the earthbound a taste of heaven. "Good and evil are typified by light and darkness," wrote Shaker eldress Aurelia Mace. "Therefore, if we bring light into a dark room, the darkness disappears, and inasmuch as a soul is filled with good, evil will disappear."

To create the heaven on earth that would attract the converts the celibate sect needed to continue, the Shakers set about expressing their faith in tangible form. In the dark, wooded Berkshires, along the muddy, rutted main road at Hancock, Massachusetts, they erected one of their farming communities, full of light, air, and space for man and beast alike. Convinced that the well-treated animal would yield more, they spent the then enormous sum of $10,000 on the dairy barn alone, a handsome round stone structure lit with three tiers of windows that glows like a cathedral even on a drizzly day. Even their chickens enjoyed heavenly light, and passed the winters cozily in a coop warmed by an early version of passive solar heating.

If the livestock got plenty of sun and air, their masters were flooded with it—a very unusual notion in the mid-nineteenth century. One Shaker describes the typical contemporary rural New England home this way: "The low, dark, and heathenishly ornamented structures are not compatible with the liberal and enlightened spirit of modern times." In contrast, a visitor described Shaker buildings as "a combination of space, beauty, symmetry, and the light and splendor of a summer's day." To attain that light, Hancock's main communal dwelling was designed with ninety-five extra-large windows of twenty-four panes each; including interior windows that "borrowed" light from adjacent rooms for central hallways, the building incorporates 3,194 pieces of glass. On entering a Shaker meeting hall, brilliantly lit, painted white and trimmed with heavenly blue, the painter Charles Sheeler noted that "instinctively one takes a deep breath, as in the midst of some moving and exalted association with nature. There were no dark corners in those lives. Their religion thrived on light, in their crafts and equally with their architecture."

The question of why Mother Ann Lee, a blacksmith's daughter and factory worker born in the industrial gloom of mid-nineteenth-century Manchester, brought the Shaker movement to America and developed a spiritual philosophy embodied by light is an interesting one. After the four children of an early, forced marriage died as infants, Lee turned inward. Following this period of withdrawal, she announced that it had been revealed to her that God was mother as well as father; although she never directly claimed to be a female Christ, she was regarded as such by many followers. Mother Ann's protracted period of grief and introspection, followed by her sense of being singled out, her inventiveness and energy, and her trances, ecstasies, and revelations all suggest that like so many leaders and creative people, she might have suffered from a mood disorder. Her search for light could have been an intuitive effort to treat the darkness of the soul.

In the years before winter depression was reidentified, many modern victims instinctively sought out bright interiors, according to Norman Rosenthal. While one woman haunted supermarkets at night, another was drawn to sit in front of her plant lights. "Vulnerable people should be careful where they live and work," he says. "Whenever possible, add lamps and fixtures, trim the foliage around windows, install skylights, remove dark wall surfaces, and use light colors in your decor. You have to be especially mindful when house hunting. A place in the Vermont woods may seem cheerful in July, but what will it be like in December, when the trees block what little winter light there is? Even in the same area, different buildings and spaces can have very different lighting conditions. One man, used to living in a sunny high-rise apartment, got a big promotion, bought his own brownstone, and became depressed. His new home may have been more elegant, but it was also a lot darker."

Going about our daily business, most of us are unaware of the drastic differences in light intensity we experience, particularly the pockets of gloom that even the sunniest climes can harbor. Another of Rosenthal's patients, a photographer, finally realized that his depressions were rooted in the long hours he spent in the darkroom, but situations far less extreme can exact a toll. "It's amazing how dim interiors can be, even next to windows," says Daniel Kripke. "And in the same house, rooms that seem virtually identical can differ greatly in terms of light. For example, people who sleep in bedrooms that have windows that face East are apt to get up earlier and sleep less than those in bedrooms with Western exposure." Becoming more aware of light's importance to the quality of life, architects and designers increasingly try to incorporate it into their schemes, and not only the rich and famous benefit. The pragmatic New York Board of Education, for example, has begun to build classrooms with a new shape. Rather than the traditional box, this version's plan resembles a bisected square whose halves have been pushed in opposite directions; because it has eight corners and walls instead of four, the room allows for bay windows and a lot more light.

Considering how much has been learned about the beneficial effects of light, it is surprising that so little is known about exactly what kind is best. About all that has been established is that brighter is better. There are no scientific data, for example, to support the popular conviction that fluorescent lights cause everything from headaches to bad moods. According to Rosenthal, the bottom line on light is that "it's the amount, not the type, that counts." Although he and his colleagues can't prescribe different types of light for different effects, they'll soon be able to offer patients more sophisticated and generally useful architectural variations on the original medicinal-looking phototherapy panels. One version, for example, looks like a window and can be used for normal illumination after close-range treatment is done; because light doesn't vary linearly with distance, if you double your distance from the source, you won't get half as much light, but perhaps only a tenth. Such devices are likely to become common appliances as more people become aware of the fact that, as Michael Terman puts it, "depression caused by light deprivation can affect the homebound, the hospitalized, and submariners just like a winter spent in Alaska."

Some people create their own light-deficient settings. While most of us gravitate toward light, a few prefer to huddle in dim rooms under the glow of a single bulb. Michael Terman suggests several reasons for this seemingly gloomy predilection. "People vulnerable to depression often put themselves in dark places—it's part of the withdrawal characteristic of the disorder. Photophobics, on the other hand, suffer from ocular problems that make bright light painfully glaring. But most people who go around turning off lamps probably do so because interior lighting is often so poorly managed. I suppose the ultimate example is the supermarket."

Scientists haven't devoted much research to supermarkets, but because of the unique psychological and social problems they pose, the environment of submarines has inspired a good bit of study. In fact, one of the few observations about the effects of color on behavior that has some scientific credibility resulted from efforts to find a soothing shade to relieve stress on subs. Named for two naval officers, "Baker-Miller pink" seems to have a calming effect, even when tested in prison cells; similar claims have been made for a blue tone that is actually a juxtaposition of turquoise and royal.

The idea of linking color and behavior is reasonable enough. Anyone who has ever felt blue, seen red, blacked out, or turned green knows we're prone to make emotional associations with different shades. We respond to colors physiologically—our eyes physically respond differently to different colors, as to different light conditions—and those who believe we react psychologically as well generally claim that "warm" ones, such as reds, yellows, and oranges, stimulate us, even to the point of inducing aggression. From this point of view, red is ideal for a slinky evening dress or a fire engine, but wrong for subway-station walls, where it could further stir up already vexed passengers waiting for a late train. On the other hand, "cool" colors such as blue and green are thought to calm the nerves, while the too-cool gray, black, and white fashionable of late are so understimulating that they can invite depression. Thus ships and planes are often painted in warm tones to counter the blues of sky and sea, while decor in deserts and other hot places often features greens. Despite extravagant claims made by "scientific" color-minded designers—say, that extroverts need cool shades to calm them down, while introverts require energizing warm ones—there is little solid research to support the notion of "right" colors for certain people or activities.

Just as the Eskimo culture developed ways to cope with the dramatic light changes of its extraordinary natural setting, technological society is beginning to learn how to adapt to the settings of our indoor, urbanized environment. The ingenuity behind noontime walking regimens and phototherapy devices springs from a strong selective pressure throughout evolution that has kept our species from becoming too dependent on a certain kind of surroundings. "One reason it's so hard to pin human behavior solely on genes is that the programmed organism dies out as soon as the environment changes," says Myron Hofer. "Think of the dinosaur. Homo sapiens has been successful in exploiting many different environments and adjusting to change because we have an open program. It's adaptive for us to be unstable, prone to make choices rather than just react to the environment."

3

.

NORTH VERSUS
SOUTH

A FTER A long, bitter controversy, Leonard Jeffries, a tenured professor of black studies at the City College of New York, was finally removed from his position as chairman of the department in the winter of 1992. Among the reasons were that he had taught that "sun people" from Africa had created a culture that was "communal, cooperative, and collective," while "ice people" from harsh European climes produced a society of "domination, destruction, and death." Despite the consternation he caused in modern academic circles, Jeffries is hardly the first to assert that a people's environment, particularly its climate, shapes their character and behavior. Generations of white racists used the same argument to explain why Africans were particularly suited for servitude on southern plantations.

Long before racism cast its dark shadow over America, intellectuals from many cultures had reasoned that peculiarly fortuitous geoclimatic conditions could foster superior citizens who were brighter and braver, more industrious and virtuous than those of other nations. Yet, like Jeffries and the defenders of American slavery, Aristotle, the Roman architect Vitruvius, and the Arab historian Ibn Khaldun each insisted that his own people's homeland was uniquely blessed, and others' benighted. As far as the Greeks were concerned, their version of the four seasons provided ideal exposure to the four elements, which helped balance the all-important four humors that determined the fitness of mind and body. The Romans borrowed this theory, applying it to their own city. To Vitruvius' way of thinking, Rome's political supremacy was proof of its environmental superiority, which was sometimes augmented by its ingenious citizens; among the many portents of the empire's fall was the decay of the capital's central heating system.

Shifting with the balance of power, the climatic ideal eventually rose from the Mediterranean to cooler Western Europe. In the eighteenth century, the political philosopher Montesquieu observed that Paris was the best of all places; people who lived farther from a coast could not be as clever, he reasoned, because their intellectual horizons were likely to be limited by mountains. Henry Buckle, an influential nineteenth-century English gentleman traveler, argued that because extremes of heat and cold interfered with work, an invigorating temperate climate was best; after all, he pointed out, the greater productivity of the laborers in such places supported his own leisure class, which immeasurably advanced the whole society. In this century, pointing to maps on which data on productivity, suicide, and even library use had been plotted, American geographer and explorer Ellsworth Huntington advanced the view that places that had four distinct seasonal changes challenged a people to be ingenious and creative.

The flaw shared by all these theories, whether formulated by Aristotle, Buckle, or Jeffries, rests on a misunderstanding of the influence of environment on behavior. How a people responds over centuries to limits imposed by temperature, humidity, rain, or wind is only one influence on their culture, which is the real architect of "national character." Like some corrupt, violent, environmentally rapacious, and backward Central American nations, Costa Rica is a tropical country. Yet this good-natured, literate, but by no means wealthy nation, a refuge for sun-loving Quaker pacifists, is led by a president honored with the Nobel peace prize. The North—the industrialized world—has about a fifth of the planet's population and uses four-fifths of its resources, but it high-handedly blames the South—the developing world—for many serious global ecological problems. Yet Costa Rica is among the most advanced nations in the world when it comes to environmental policy; for example, it has conserved 26 percent of its wild lands, as opposed to America's 5 percent, and exposes every schoolchild to nature study. The climate Costa Rica shares with its neighbors could not be responsible for its own enlightenment or their plight. Its progressive culture derives from the fact that it was colonized by rank-and-file Spanish soldiers rather than the grandees who imposed the terrible tradition of rich landowner and poor peasant that continues to oppress other countries in that part of the world.

"You hear these North versus South stereotypes—that the environment 'makes' northern people hard-working, intelligent, and upright and southern ones lazy, dull-witted, and amoral," says Kerry Feldman, who has done research in Philippine squatter settlements as well as in Alaskan whaling villages. "You can't say that environment determines 'northern' or 'southern' behavior, because a constant can never explain a variable. The hot regions gave us our first civilizations and some of the most creative, industrious ones the world has ever seen—think of the Mayan architects, the mathematicians of the Arabian peninsula. When Westerners speak of the Middle Ages, we mean Europe, but there was an equally magnificent medieval period in India. Great civilizations couldn't be determined by climate or genes, because they spring up in such different places—Egypt, Germany, China—and then die out. The Greeks and Romans gave us Western civilization, but the Mediterranean area is considered backward now. Who wants to know the latest Greek theory about anything? The era of Euro-American dominance has lasted four hundred years, but a century from now, people down in Brazil might be saying that North Americans are lazy and shiftless."

During the summer of 1988, Brazilians might have been amused by a *New York Times* headline that, with an uncharacteristic hint of hysteria, noted that the metropolitan area was embarking on its sixth week of a heat wave that had produced the same atmospheric conditions that prevail in the Amazon basin. The weather may have been Brazilian, but New Yorkers' responses were not. With varying degrees of success, they tried to carry on in the best mad-dogs tradition, maintaining the kind of activity level that precipitates a thousand deaths in the fifteen largest U.S. cities each summer. These fatalities caused by health problems aggravated by heat occur at lower temperatures in northern cities than in southern ones; in New York, the incidence climbs at 92 degrees, versus 103 in Dallas. As this contrast in the precipitating temperatures of heat-related deaths makes clear, some of us, mainly southerners, handle sweltering weather better than others, and in the old days, this difference was said to be "in the blood." Modern research, however, indicates that the resilience of people from the hot parts of the globe depends far more on behavior modification than on biological adaptation, even among groups who seem physically well-suited to their extreme environments. Although the Tuareg of the Sahara have tall, lean bodies that produce and retain less heat than stocky ones, their chief means of dealing with the desert are loose garments that block the sun and a tradition of taking it easy during the hottest part of the day.

No matter where we are from, we can physiologically adapt to the Sahara or the Amazon basin in a few weeks, mostly because our bodies will "learn" to sweat sooner and more efficiently. Far more important, if we are clever we will quickly acquire the siesta mentality. Unlike Yankees, people from hot places tend to work early or late in the day, rest in the mid-afternoon, and generally proceed at a more moderate pace. Whatever laziness is associated with hot climates, particularly humid ones, is likelier to be caused by sicknesses that flourish in those conditions, especially those carried by parasites, than by temperature.

Most of our species' ability to cope with extreme cold also depends on behavioral modification. All Alaskans—Native, white, and black—face the stress of winter cold so extreme that when they take out the trash, the hairs inside their noses freeze, but any Alaskan can tell you who wears the chilly mantle of winter most lightly. "I've never heard an Eskimo complain about the temperature," says Jerry Mohatt in Fairbanks. "They never even bring up the weather in conversation. It can be sixty below, and they walk around with their hoods down. They just don't seem to have the same perception of cold."

It makes sense that they shouldn't—so much sense that the theory is included in medical texts. "According to most physiology books, the southern Negro is likeliest to suffer from freezing," says William Mills, the cold-injury specialist in Anchorage. "The southern white comes next, followed by the northern Negro, the northern white, and last, the Eskimo—presumably the one best conditioned for his native climate. But my experience here in Alaska has been that Indian and Eskimo, black and white all freeze equally. Most Natives I've treated have told me, 'I didn't realize I had frozen feet until I took off my mukluks.' One reason they don't feel the signs of freezing as soon is that they've acquired the ability to keep feeding warm blood to their extremities. If you've had no exposure to cold, your vessels do the opposite and contract. After about six weeks, however, a newcomer would also develop increased circulation in the hands and feet."

Whether Eskimos, the Indians of Tierra del Fuego, and the bushmen of the Kalahari have extra genetic help in adapting to cold is unproved. Contrary to what we may have learned in school, however, Eskimos aren't born insulated with extra body fat; in fact, the restricted diet of a subsistence life-style inclines those who follow it to be lean. Nor does the shape of their eyes offer adequate protection against glare and snowblindness; Eskimos are vulnerable to eye problems and often use goggles or visors. In short, even in their extreme environment, the Inuit show only a few physiological adaptations to cold. Along with the peripheral blood flow Mills describes, they may have a somewhat higher metabolic rate, perhaps because they must spend twice as much energy on their outside activities as we do in temperate environs.

Lab experiments have shown that while a rat can adapt to cold in a place where it has learned to expect it, it can't maintain the adjustment in a new environment of the same temperature. William Mills suspects something of the sort applies to people as well: "If you brought up an Eskimo baby from birth in New York, then took him back to Alaska twenty years later, my gut feeling is that his ability to tolerate cold would be diminished." Like the rest of us, the young Eskimo could physiologically acclimate to Arctic conditions in a few weeks, but most of his adjustment would involve learning how to pile on the right clothing and seek the right shelter, by far the most important ways people cope with cold.

Even what we think of as the objective reality of our native environment has a strong learned component. When a pygmy who had lived his whole life in the dense African forest was taken for the first time to an open area, he completely misjudged the sizes and relative distances of unfamiliar things, perceiving distant cattle as nearby insects. Similarly, studies show that when confronted with a short line drawn on a piece of paper, subjects reared in vast open stretches of desert or tundra interpret it as a long line that extends back into the distance. Feldman offers an Alaskan illustration of how deeply early perceptions of the environment are etched into our brains: "Knowing that his people's word for tree refers to a spruce, the only tree they have, I asked my Inupiat Eskimo friend Ernie, who has a degree in anthropology, how he perceives a birch tree. He said, 'I experience "big bush." I said, 'Ernie, you've been in our system long enough to know we experience that as "tree." 'I know,' he said. 'I still remember how my brain hurt when I was six years old and the white teacher told me that *aaglu* was a whale. How could you people possibly think *aaglu* is a kind of *agvik*? *Aaglu*, what you call the killer whale, is black and white. It's got a dorsal fin and teeth. It's meaner than hell, and hunts like a human. My brain hurt.' "

If we have one gauge of what is real, it is seeing something "with our own eyes." But Feldman points out that even this measure can be far from objective. "Some Ethiopian women were asked by Western researchers to sort colored disks into piles," he says. "Now these ladies had worked as maids for Europeans in the nearby town, so when they got to the blue and green disks, they said, 'Excuse us, but do you want us to do it the way the Europeans would and put them into separate piles, or the way we see it, putting them together?' Along the Mediterranean and in other places where intense sunlight has caused dark eye pigmentation, people often have a single word for blue and green, which are very close on the spectrum. A brown-eyed person may never know what color means to a blue-eyed person. Like the Ethiopian women, they probably see the difference, but it's just not an important part of their reality. One group in the Philippines has to know whether, say, a leaf is wet-green or dry-green before they can classify its color. We assume that perception influences cognition, as if we were rational machines that process incoming data and spit out objective maps. But from birth, the way our culture teaches us to think constantly filters what we call reality. Snow is one thing to us and another to the Inupiat, who have thirty different terms for it."

Yet despite perceptual templates laid down as far away as the tropics, newcomers to Alaska, where nearly forty languages are spoken, not only adapt to the state's environmental extremes but prosper. "There are three hundred Filipinos, straight from their homeland, living in

Barrow right now," says Feldman. "Eskimos traditionally look down upon cleaning hotels and serving food—no fast-fry *maktak* [whale blubber] stands. So Filipinos have come to Alaska to do what they do back home if they don't have much money. They go into the service sector. You could take Eskimo people from here and move them to the Philippines and they'd adapt, too. That's what human beings do. When I did a study of whale hunting in a small Arctic village, one of my key sources was a Hawaiian Chinese man. He had married an Eskimo woman he met at college and moved to this little settlement. At first I thought he *was* Eskimo. So here's this Hawaiian Chinese man in Alaska, and before you know it, he's out hunting whales. Culturally, he's different, but biologically, he can do whatever the Eskimo can. His children are going to be Chinese-Hawaiian Eskimos, and that's the future of the world."

Climate may not determine behavior or culture, but it can influence them by setting up certain limitations. Anyone who has survived the dog days of August to wake up on the first crisp September morning has a gut-level understanding of science's venerable concept of how temperature affects behavior, which has traditionally been exploited by doctors to help their patients. "Emil Kraepelin, the great nineteenth-century theoretician of depressive disorders, used heat to calm victims of mania," says Thomas Wehr. "After I gave a talk at a very old mental hospital in North Carolina a while ago, I learned that its depressed patients used to be sent home with the advice to take cold baths."

There are no statistical records of how well these simple thermal therapies worked, but an engineer would find the principles behind them to be sound. Most of our physiological adaptations to heat and cold depend on the cardiovascular system. When we are too warm, it sends blood to the surface in order to cool the body's core, and when we are chilly, it does the opposite, constricting vessels and keeping blood near the internal organs. Because our muscles generate 80 percent of the body's heat, maintaining the right temperature no matter what is going on outdoors depends a good bit on how active or inactive those muscles are, which in turn depends a good bit on our state of mind.

Cold affects us like a stimulant, inspiring us to get our heat-producing muscles moving. We stamp our feet and rub our arms to warm up at a football game, and that kind of energetic response to cold helps ensure survival in more extreme circumstances. "Suppose your small aircraft goes down or your dog team runs away, leaving you right in the middle of stark raving tundra, with nothing but ice and snow as far as you can see, and maybe severe wind," says Mills. "In that kind of situation, you better not just sit down and curl up. You better get busy."

Just as cold is a stimulant, heat is a sedative. If we bustle around on a hot day and get overheated, our instinct is to switch off our muscles, rest, and cool off. "To decrease your body's heat production, you have to stop activity," says Wehr. "To do that, you have to become disinterested, and to do that, you have to think there's no point in doing things. Over time, that's depression." According to his elegant but unproved hypothesis about the roots of mood disorders, depression is a caricature of our normal slow-down response to heat, and mania a caricature of our get-busy reaction to cold. "At this point, we have nowhere near the amount of evidence for this mechanism that we have for the role of light, and there are many other influences on those states," he says. "But the concept helps us begin to think about the role of temperature on behavior."

Some eclectic studies conducted in places as diverse as schoolrooms and battlefields support Wehr's thinking about heat. During the Persian Gulf war, American troops suddenly found themselves sweltering in daytime temperatures of up to 130 degrees. Even sailors diving off battleships for a swim under the watchful eyes of riflemen alert for sharks splashed into 100-degree water. As many as 25 percent of soldiers from the Northern Hemisphere who are suddenly landed in such tropical conditions are unfit for combat. Even in far more temperate settings than those of the Middle East or Southeast Asia, an hour's exposure to temperatures over 90 degrees tends to impair physical performance, while two hours' worth interferes with difficult mental tasks. Not everyone responds to heat this way, nor do all types of effort diminish equally, but generally, the more the temperature rises over 90, the faster these declines set in. Aware of the debilitating effects of heat on behavior, military psychologists in the Gulf were ready to test a new kind of hot-weather pick-me-up: a specially designed vest whose pockets hold twelve pounds of ice packs. Soldier-subjects who wore it reported that their problems with sleep and poor spirits decreased noticeably.

The soldiers' reaction to relief from the Middle East heat resonates with the experience of a woman known in psychiatric literature as Marge. During the warm, sunny months of the year, she felt like herself only while on vacation, swimming in the chilly Finger Lakes of New York. Once she was back home in steamy Washington, D.C., her mood and energy level plummeted, briefly rising during the occasional July or August cool snap. Marge assumed she felt better up North because she was at leisure, not because she was living in a different physical environment. In the course of seeking help for her recurrrent distress, she was finally referred to Wehr, who had begun to monitor a less common seasonal syndrome that seemed to be the reverse of winter depression.

Following newspaper accounts of NIMH research on the February blues, calls and letters describing a different sort of seasonal complaint had started filtering in to the institute's campus in Bethesda, Maryland. Some were from local people who said they grew depressed during the area's hot, muggy summer and others from globe-trotting politicians and diplomats stationed in the capital who claimed that their various postings had profound effects on their well-being. "As soon as they got off the plane in, say, Guam, and felt that humidity, they said they knew it was going to be a very bad year," says Wehr. "We forget that the humidity level, which is the amount of water in the air, changes just as temperature does in summer and winter. In Guam or right here in summer, each time a water molecule evaporates from your skin, another one lands, sealing in your body heat under a layer of wet insulation."

Like Marge, Wehr's other summer patients reported that they felt poorly from June till November, just when natural light was most abundant and his winter SAD patients felt best. There were some other differences between the two groups as well. Those who slumped in summer tended to eat and sleep less rather than more, as many winter depressives do, and to endure spells of anxiety, even in the virulent form of panic attacks, which Wehr began to suspect were also bound up with heat. What the summer and winter patients had in common was depression—chronic feelings of helplessness, hopelessness, and inertia—that came at one time of the year and left at another. Like most people troubled by seasonal shifts in behavior, they had never connected what was going on in the outside world with what was happening inside themselves. Acting on a hunch, Wehr suggested that Marge while away a week of Washington's summer within the boundaries of an air-conditioned environment. As the mercury fell indoors, her depression lifted. "One of our other patients turns into a kind of mole in the summer, moving from his air-conditioned house and garage to his air-conditioned car and office," says Wehr. "That clamps down on temperature pretty well, but there's a limit to how much humidity, which also carries the heat signal, you can get rid of in a place like Washington."

Just as winter's darkness depresses some people, Wehr and his colleagues think summer's heat cripples others. Although set in motion by different cues from the sun, both types of seasonal melancholy probably make trouble somewhere in the hypothalamus, the brain's Grand Central Terminal of behavior. Right beside the entry point for incoming information about external light conditions sits the body's thermostat, which senses external thermal conditions and triggers appropriate responses, from shivering to sweating. "Just as the thermostat in your hallway coordinates the temperature of your house to the weather, the hypothalamus helps your body adjust to what's going on outside," says Norman Rosenthal. "Winter patients seem to have a problem in the neural connections between the hypothalamus's light sensor and the eye. Summer patients may have trouble with the way heat signals are received and responded to by its thermal regulator. A smaller group of people who get depressed in both winter and summer appear to have difficulty in processing both heat and light signals, and only feel well in the spring and fall."

Just as most people's spirits sag a bit in the darkest months of winter, says Michael Terman, "the population as a whole slumps in July and August and rebounds in September and October." Although surveys of seasonal mood swings done in different latitudes show that residents of, say, Florida are much likelier to feel poorly in summer than New Englanders do, not all researchers are convinced that this dip in spirits is necessarily provoked by heat, however, or even that there's a distinct form of summer SAD. "It's possible that some people stay indoors to escape the hot weather and become light-deprived—that can happen at any time of the year," says Daniel Kripke. "The problem isn't the weather, but their reaction to it."

Long before Wehr began prescribing air conditioners, some people upset by summer had come up with an environmental treatment of their own: head north. Finding the ideal place to conduct research on the treatment of cold injuries was not the only consideration that enticed Mills to Alaska. "If I travel too far down into the Lower Forty-eight in summer, I'm absolutely miserable," he says. "I'm a winter person who finds cold exhilarating. I love all the skiing, skating, hockey, and ice-fishing we do up here during the coldest months. I actually look forward to the moment every Fourth of July when my wife says, 'It will snow soon.' "

In the event of a "global warming" of several degrees, caused by heat trapped in the atmosphere by pollutants, some environmentalists warn, the ranks of Alaskan immigrants might swell. However, Paul Bell, a professor of psychology at Colorado State University who has conducted research on the effects of heat on behavior, isn't concerned about such relatively minor temperature changes. He points out that the stuff the world's cities are made of and the technology they employ have already made them about ten degrees warmer than the surrounding countryside, and that "on any given day of the year, you can find variations of thirty to fifty degrees in the same location. Changes caused by global warming would pale in comparison to those nature regularly brings about."

As Bell observes, we don't have to change latitudes or seasons to experience significant thermal differences. In the spring or fall, when we leave for work in the morning the temperature might be 40 degrees, yet climb to 75 by lunchtime. Like light, the external temperature changes from day to night, helping us keep in sync with the sun's cycle. And like our built-in responses to changes in light, our own bodies' thermal ebbs and flows are probably rooted far back in the days and nights of our evolutionary past. The linkage Wehr makes between temperature and mood rests in part on his observation that sleep, like heat, is a depressant. When we doze off, our internal thermostat sets down, which eventually lowers our temperature; however, the body initially reacts to this adjustment by feeling too warm, which triggers cool-down responses such as sweating. On the other hand, like cold, wakefulness is a stimulant.

It is not possible simply to say that mood is a function of temperature, because in normal people it isn't markedly affected by the daily rise and fall of the mercury or the internal chemical changes associated with sleep and wakefulness. Yet among those prone to mood illness, there are some intriguing links between state and thermal regulation. For example, Sally Severino, a researcher in chronobiology at New York Hospital–Cornell Medical Center, found that the nocturnal temperatures of women prone to PMS are significantly higher than those of other women. And Wehr finds that 60 percent of his depressed patients cheer up when they stay awake all night, then relapse when they finally go to sleep; manic-depressives go past happy to hyper when they stay up, which further feeds their insomnia. One woman illustrates both these phenomena: after being kept awake all night to alleviate her depression, she became manic. Following two days of wakefulness, she finally dozed for about forty-five minutes—and relapsed into depression. "Perhaps," says Wehr, "some people prone to depression don't regulate their heat very well late at night, when mammals who are active by day are supposed to have lower body temperatures and rest quietly in their burrows."

In July 1991, on the 100-degree-plus day on which its fans and air conditioners set a record for the most electricity New York had ever used, the city came close to breaking another, logging the second-highest daily tally of murders in its history. Noting other unhappy developments on the overheated urban scene, from increased racial tensions to child abuse, newscasters made the references to the "long hot summer" that have become traditional since a report by the U.S. Riot Commission observed that all but one of the infamous 1967 "race riots" began on warm days.

Although temperature alone can't make us act up, lots of facts and figures connect soaring mercury and bad vibrations. The incidence of assaults and murders is highest in summer, more violence occurs in the southern regions of countries in the Northern Hemisphere, and even the FBI lists hot weather as an important consideration in the epidemiology of crime. Not that we have to be material for the Most Wanted List to get ornery: drivers of cars without air conditioning honk their horns more when the temperature is over 85 degrees, and on the baseball field, batters are likelier to be hit by wild pitches on such days. The summer surge in bad behavior has traditionally been attributed to social factors, such as the fact that people have more free time to get into trouble, are often outside,

wear fewer clothes, and may drink and use other drugs more frequently. However, there is evidence that while heat cannot cause us
to pull the trigger or throw a dish, it can exacerbate other stresses
to make such behavior more likely.

Several studies have challenged the social explanation of the long-
hot-summer phenomenon. In one analysis, a team of researchers
found that the incidence of wife-beating peaked in the hot months
and slacked off in the fall, when, according to the cultural hypothesis,
it should increase because people are cooped up indoors; moreover,
the incidence of abuse was highest in the hottest places. Another
common form of violence, suicide, also has a season, although the
long-recorded variations in the incidence of suicide have decreased,
especially in urban areas, since the advent of artificial heating, cool-
ing, and light. "Suicide peaks earlier than spouse abuse, in May and
June, because unlike battering, it can't be repeated," says Wehr. "You
can't start to kill yourself in May and finish up in August—your first
time out is it."

While a good deal of research indicates that the hotter it gets, the
higher the incidence of aggression, Bell has found that above 92
degrees, people reacted less aggressively to provocation, preferring to
avoid or ignore it; their response suggested that when they felt way
too hot, his subjects tended toward "flight" responses rather than
"fight" ones. "Generally, as the temperature increases, there's more
violence," he says, "but the higher it climbs, the more variation you
see. On the day when New York had all those murders, some other
place in the country that was just as hot surely recorded a very low
incidence of crime. In the real world, it's hard to isolate heat as the
factor in aggression. We like simple explanations for behavior and
don't want to consider three or four variables, much less a hundred."

In some parts of the world, wind is among the variables thought
to intensify the aggravating effects of heat. When the hot, dry Sharav
blows across the Middle East, for example, some judges regard impulsive crimes more leniently. Appearing at particular times of the
year, such regional winds, including the chinook of Colorado and
Wyoming and the Santa Ana of California as well as the Swiss foehn,
have been blamed for depression, irritability, poor performance,
aches and pains, and accident-proneness, as well as crime.

Just as he suspects that heat promotes lethargy to cool us off, Wehr
has a commonsensical illustration of why it makes us unsociable.

"There's a series of textbook pictures, taken from above, of a whole bunch of piglets at different ambient temperatures," he says. "As it grew colder, the piglets got closer and closer until they ended up in a little pile in the middle. When it got warmer, they dispersed more and more, until at the highest temperature, they were separated from each other to the maximum degree allowed by the stall. Maybe when you get too warm, you just want others to keep their distance so you can cool off."

One time-honored, peaceable solution to dealing with the increased arousal caused by heat has been to make love instead of war. Extensive records imply that temperature fans the flames of romantic as well as violent passions. "All over the world, conception has peaked in the late spring in a big way and in a more minor way in early fall," says Wehr. "Then, several decades ago, in what's known as the American pattern, the spring peak diminished and the fall one increased. We've shifted major environmental influences on behavior, such as heat and cold and humidity, in a very significant way with central heating and air conditioning, and the rest of the world is catching up to us fast." Bell leans toward a cultural etiology for warm-weather mating: "More people are on vacation."

The extraordinary tradition of helpfulness to be found year-round in Alaska is a powerful testimony to the importance of staying warm in a cold place. Just as we tend to go off alone to "cool off" a hot temper and try to "stay cool" during the long hot summer, we may be inclined to be more sociable when we need to keep warm. "Here in Fairbanks, if a family gets into difficulty—say, their house burns down—everybody just pitches in," says Rothrock. "But people talk wistfully about the time before our big growth spurt, when doors were never locked because somebody might want to come in and warm up while you were gone." Even in larger, more cosmopolitan Anchorage, says Feldman, "although there are people who like their BMWs and Jaguars and art auctions, that doesn't mean much in the middle of winter. Winter is a democratizing context. If your car doesn't start, you need someone to help you, and someone will. Status means nothing if your car won't start."

Gripes about temperature are our commonest environmental complaint. Although most of us find 71 degrees close to ideal, 10 percent consider that either too hot or cold. To stay alive, however, we must keep our body temperature below 113 degrees and above 77; to feel and function well, we need to stay near 98.6. When creatures are unable to keep warm, "there simply *isn't* much behavior," says Paul MacLean. "For a long time ethologists didn't discuss reptiles very often. The animals were observed at room temperature, which is too cold for them, so there wasn't a lot to say. Our little squirrel monkeys here at NIMH won't even breed unless the temperature and humidity are right. Without heat, we living things are just nothing. It's important to everything we do."

Although we are able to tolerate colder places than MacLean's lizards can, when the mercury drops below freezing and we are unprepared, we too react like reptiles. In fact, a predictable series of behavioral changes shows just how close a victim of hypothermia is to freezing to death. "By the time his body temperature has fallen to about ninety-four degrees, he speaks as if drunk, with a mouth full of mush—the cardinal sign of being in a lot of trouble," says Mills. "Then he starts falling and loses visual acuity—certainly he's no longer capable of functioning well. By ninety-two degrees, he pays no attention to where he's going or what he's doing. For some reason, he suddenly begins shedding his clothing. When he's finally found, you will see shoes and socks on the ground. A woman will have her dress up and panties down, a man will have pulled his shirt up and dropped his pants. Perhaps the person experiences a sensation of flush in the groin—the body's last effort to develop heat. This 'paradoxical undressing' presents an interesting problem for us up here, because when the police come along, they often figure the victim has been sexually assaulted. If they cordon off the area because they think they have a dead crime victim, we don't get the chance to resuscitate."

Because cold so effectively knocks out behavior and even the appearance of physiological functioning, victims of hypothermia can be mistakenly declared dead. "Very few physicians and emergency people keep the stethoscope on the chest for two or three minutes, but if they do, they may find a little blip that's evidence of life," says Mills. "If it's fifty below, and so is my victim, he's dead. But if it's fifty below and his temperature is seventy, I just have to assume life. I wait until he's warm back at the hospital and if he's still dead, then I say, you know, he's dead."

4

.

GOING TO
EXTREMES

"Extreme environment" tends to be a relative term, but anyone would apply it to Mount Denali, the highest mountain in North America and the coldest in the world. Although many places labeled extreme seem perfectly ordinary to Bedouins, Tibetans, or Eskimos, the locales that usually warrant that description are very hot, cold, high, dry, wet, or windy. Some, like Denali, are a combination. During his ascent, one climber began to hyperventilate and go numb. His more experienced companion jumped on him, yelling "Stop breathing!" Aware of the hazards posed by the deadly combination of high altitude and severe cold, he knew the novice was inhaling frigid air too quickly, thus freezing himself to death from the inside out. Along with whatever physiological stresses such extraordinary places might impose, they are usually remote, potentially dangerous, alternately nerve-racking and boring, and apt to require either solitude or monotonous involvement with a small group. For all these reasons, extreme environments intrigue scientists who study the psychological and physiological responses of different people to challenging conditions and the problems of living in previously unthinkable places, such as under the sea or in space. Although conducted in settings far beyond our wildest dreams—unless we are enthusiasts of increasingly popular high-risk sports and wilderness experiences—research on extreme environments is producing important insights into how our species adapts to everyday ones as well. Some of the most interesting and widely applicable studies concern the high-altitude milieu, which affects not only mountaineers but also, say, the millions of unsuspecting tourists who drive through the Rockies on vacation.

Each summer, Peter Hackett, a physician and leading researcher on the physiological changes wrought by altitude, spends two months working at a lab halfway between heaven and earth. Some of his six team members take five to seven days to climb to the University of Alaska facility, plunked on a large glacier 14,000 feet up on Denali, about 6,000 feet below the summit. Although he has done that many times, Hackett, a world-class climber who has also scaled Everest twice, usually helicopters to the mountain shelf. The cluster of buildings—plastic shells of 12 by 14 feet supported by steel frames, floored with plywood, and heated by propane—is shielded from avalanche by a kind of natural amphitheater. Roped to prevent themselves from tumbling into a crevasse, team members can walk about 200 meters in almost any direction.

Hackett sums up the general ambience: "Because the lab sort of mimics conditions on the space shuttle, NASA has used it to study the dynamics of a small isolated technical group with a mission to do in a hostile environment. You are very much at the mercy of the elements up there. You can get depressed after four or five days of hundred-mile-an-hour winds that keep you from going out and doing anything. There's nothing more unnerving than trying to get a little sleep inside your tent with your boots on and your ice ax next to you—if the tent explodes in the wind, you want to be prepared to smash it into the snow immediately so that you don't get blown away. Those kinds of stresses can be very trying. But at other times, looking out over the clouds and the mountain ranges, you can feel quite euphoric about where you are. What's most impressive is the absolute lack of most life forms—no plants, insects, or mammals, and birds rarely. There's only ice and rock, yet there are moments of intense appreciation of the raw beauty of this austere environment."

Hackett more or less backed into a special vocation that combines one part researcher, one part doctor, and one part Indiana Jones. "After my internship, I wanted a break to see what the real world was like before going on to study psychiatry," he says. "I had already done some climbing and worked as a helicopter rescue doctor in Yosemite, so I ended up taking a job accompanying trekking groups to Nepal. I just fell in love with the mountains and the people." In the Himalayas, Hackett was called upon to treat a lot of altitude sickness. When he reviewed the scientific literature, he found there wasn't very much. Over the past fifteen years, he has written some fifty papers on hypoxia, a low level of oxygen in the blood that affects victims of certain lung and heart diseases at sea level, as well as climbers and mountain populations. While working on frigid Denali, he has also become interested in the effects of extreme cold. When not doing his research, Hackett is busy in the emergency room at Humana Hospital in Anchorage, treating the survivors of big-game-hunting accidents, small plane crashes, and frostbite.

Hackett stresses that you don't have to be a climber to feel the effects of going where the air is rare. Travelers on big jets that cruise at altitudes of 35,000 to 42,000 feet are subjected to air-pressure conditions that approximate those on mountains of about 6,000 to 8,000 feet, which means their oxygen supply is diminished by 25 to 40 percent. Each year, millions of hikers, skiers, and other visitors to Colorado, which contains about three quarters of the nation's country over 10,000 feet, have a high-altitude experience, "but most are unaware of how many ways they're affected," he says. "Anytime we're over five thousand feet, lots of things in our minds and bodies change, from an increase in the alkalinity of the blood to decreased light sensitivity in the eye."

As we ascend, what the brain cares about isn't altitude per se but the level of oxygen in the blood. Under normal barometric pressure, oxygen diffuses into the body through the alveolar walls of the lungs; the difference between the oxygen pressure of the blood and that of the external atmosphere helps force the gas into the body. When the air pressure is low, less oxygen enters the lungs, and its level in the body falls. Even at sea level, we experience changes in air pressure when the barometer falls just before and during stormy weather. In the seventeenth century, the poet John Taylore observed that "Some by painful elbow, hip, or knee/Will shrewdly guesse what weathers like to be," and certain elderly people insist they can predict rain by the creaks in their joints. Some studies have linked low-pressure changes with increased psychiatric admissions, disruption in schools, mood swings, and medical complaints, especially arthritis; like most environmental variables, barometric lows wouldn't cause such complicated problems but could aggravate them.

At higher altitude, unpleasant symptoms become commoner. In western ski areas, a surprising number of people feel so poorly that they must drop out of classes, forfeit costly lift tickets, or even stagger off to bustling infirmaries to breathe oxygen. Although acclimatization eliminates most of their distress in time, "the Colorado ski resorts report that fifteen to twenty percent of their guests—millions of people—suffer from mountain sickness," says Hackett. "Those who just have headaches, insomnia, and lack of appetite can use aspirin or just take it easy, and the problem will generally pass in twenty-four to forty-eight hours. It's the people who don't listen to their bodies and try to push themselves who get in trouble."

A big mountain is not a good place to invite trouble. Life expectancy at 20,000 feet—Denali's summit—is unknown, but above 30,000 feet, it's only two to six days. As you climb beyond a certain point, it becomes debatable whether you really remain on the earth. Slowly, you ascend to an environment like the one you cruise through on a big plane, but without the pressurized cabin. "At eighteen thousand feet, you're at half an atmosphere," says William Mills, who has treated scores of climbers for cold injury. "When air pressure is half the normal level, lots of interesting things happen. For example, let's say you're wearing boots that have a hard exterior shell and a neoprene liner. As you get to eighteen thousand feet, the lower atmospheric pressure allows the liner's cells to expand, but the shell won't give. The expansion of the lining is directed downward against the foot, cutting off circulation just like a tourniquet and causing severe freezing. And, of course, low oxygen makes you struggle twice as hard to get the same amount of air."

Above 8,000 feet, an individual's physiology or failure to acclimatize gradually and take it easy can bring on severe altitude sickness: breathlessness, weakness, giddiness, headache, insomnia, and even edema of the lungs and brain. Just as people differ in their responses to extremes of light and temperature, not everyone maintains the same level of oxygen in the blood or adapts to altitude equally well. Those who adjust easily have a highly responsive carotid body, a little gland in the carotid artery in the neck that detects the amount of oxygen in the blood and tells the brain to initiate hyperventilation to make up for a lack. "If we go up to fourteen thousand feet right now and you have a good carotid body response and I don't, our blood oxygen levels could be very different," says Hackett. "Mine might resemble that of somebody at twenty thousand feet who is doing okay, which means I can get in trouble even at five, eight, or ten thousand feet. The people who don't respond well tend to be very good endurance athletes, because they don't get breathless when exercising. Physical conditioning has nothing to do with ability to acclimatize to altitude—it's the great equalizer. Sedentary slobs do fine."

Even though the Denali team members take a drug called Diamox to help them adapt more quickly to altitude, when Hackett gets off a helicopter at 14,000 feet, he soon feels lethargic and a little dizzy. "I generally vomit a couple of times within the first couple of hours, and that night is terrible," he says. "In a few days I feel better, and eventually I feel good." Some people never adjust, however, and for them, says Hackett, "there are three cures for mountain sickness— descend, descend, and descend. There are people who've lived in the mountains all their lives but have never acclimated, and they suffer from a chronic version. To feel well, they have to move."

There is little evidence that altitude fitness is inherited. "If there are any genetic adaptations or markers for altitude, we certainly haven't identified them," says Hackett. "When we examined the Quechua Indians of the Andes and the Sherpas of the Himalayas, we found that both have acquired some physiological adaptations that distinguish them from people who live at sea level. For example, the Tibetans living at fourteen thousand feet in the Himalayas are a small people whose growth seems a bit stunted by the environment. The boys mature late, and because menarche occurs about three months later per thousand feet, the girls don't menstruate until sixteen to eighteen, instead of ten to twelve. Tibetan women are tough, and have a lot of responsibility and power. It's not a matriarchal society, but it's not really patriarchal either. If there is a divorce, for example, the property is split. The late sexual maturation caused by the environment means life there is a little different."

Like cultures that developed in extreme cold or heat, those that evolved up high rely on behavioral adaptations to their environment. Rather than trying to "subdue the earth" in the Judeo-Christian tradition, they tend to suit their ways to nature. "Real competitive jocks don't do as well up high because they're apt to go faster and push themselves more," says Hackett. "The one thing you should do in that kind of environment is reduce the tension between what's going on in the world and what's going on in the body. You need to have more breaks, walk more slowly, just take everything a little easier. It's the macho types who get in trouble."

Many claims have been made about the effects of mountains on the cultures that develop there. According to the *Book of Common Prayer*, "The mountains shall bring peace," but Peter Suedfeld, a professor of psychology at the University of British Columbia who studies the behavioral effects of extreme environments, observes that "people from the mid-level altitudes of the Alps, the Pyrenees, the Scottish Highlands, and the Appalachians share a reputation for being tough, independent, clannish, and good in a fight. I don't know how valid this theory is, but it makes sense in terms of the isolation and more difficult conditions in mountainous regions." Yet, like anthropologist Kerry Feldman, Hackett points out that culture can create exceptions to any rule. "The Sherpas generally seem happy, proud, independent, and well balanced," he says, "and it's tempting to attribute that to their incredible environment. But in very similar conditions, the Peruvians seem kind of chronically depressed. Perhaps the reason they act oppressed and crushed is because, unlike the Nepalese, they have been."

Long before honeymooners found the Poconos or British royalty took to the Scottish Highlands, mountains were regarded as places of wonder and even, in the case of Sinai, Olympus, and others, as sites for communication with the divine. Cultures that didn't have these natural status symbols sometimes built them, leaving behind the pyramids and ziggurats of the Middle East. Much of the old sense of fear and reverence inspired by mountains was transformed in the nineteenth century by the Romantic movement in the arts and the accelerated exploration of the world's wild places, which gradually changed them into havens for recreation and restoration. Yet mountains remain powerful religious and political symbols: his audience understood exactly what he meant when Dr. Martin Luther King

GOING TO EXTREMES ■ 73

thanked God for allowing him to go to the mountain, and it is no accident that in most bureaucracies, the big shots occupy the top floors of the headquarters.

The vast number of mountain resorts scattered over the earth are the best testimony to the fact that many of us feel different at moderately high altitudes. Some of that difference seems psychological— a state of energy, lightness, and good humor can even prevail over the headache and insomnia of the adjustment period. "I live not in myself, but I become/Portion of that around me; and to me/ High mountains are a feeling, but the hum/ Of human cities torture," wrote Lord Byron in "Childe Harold's Pilgrimage." Hackett agrees. "My bias is that there is something inherent in people that makes them feel better and more centered in the mountains than in other places. There are definitely mountain people and ocean people, and both those environments provoke an appreciation of natural forces at work and an energy more important than human power, and a sense of our proper place in the world. I've come to understand that there's an interaction between people and environments that changes us, and I suppose it also changes the environment. Which is the actor and which is the reactor is not always clear."

A good bit of mountain magic is aesthetic. As John Ruskin put it, "Mountains are the beginning and end of all natural scenery." The radically different scale of their landscapes lifts us out of the rut of habitual perception. "The first time you go to the Alaska range or the Himalayas, it takes weeks, even months, to get a feeling for judging distances and size," says Hackett. "People get in trouble because they decide a certain destination is only a day away. It turns out to be four or five days off, but they didn't bring any camping equipment. Different microenvironments at altitude affect your perception as well. If you're in a closed forest, you'll read things differently than you would if you were up on a ridge line, where you feel on top of things and command all views."

As a student of the external influences on internal states, Thomas Wehr takes a more pragmatic perspective. "What is it about these high-altitude places, such as the 'Magic Mountain' in Switzerland?" he muses. "You get sunlight, which we think is good. You get cold, which we think is a kind of antidepressant. Low air pressure, which reduces the amount of oxygen in the body, works like cold in that your sweat evaporates more rapidly—that's why they're always offering you drinks on planes—causing you to lose a lot more heat. The higher you go, the more altitude's effects resemble those of cold exposure. People who have used hyperbaric chambers to treat arthritis and other diseases have reported mood reactions, and we've had a couple of patients who switched out of major depressions when they flew at high altitudes. I can only speculate that their reaction has something to do with our observation that cold has antidepressant qualities and heat depressant ones, at least for certain people. Maybe altitude is a stimulant to the extent that it mimics or enhances cold exposure."

If you ask people who love mountains to explain what is so special about their favorite environment, a number will think for a minute, shrug, smile, and say, "Just something in the air." Particularly for people with certain respiratory problems, it is probably more a matter of what is *not* in the air. Because the atmosphere is much less humid and freer of allergens and pollutants, it's just plain easier to breathe up high. The air quality is so much better that most visitors never notice the impairment in their night vision caused by a low oxygen level; the stars look just as bright. Climbing scientists are apt to attribute that Rocky Mountain high to the drop in oxygen as well. "We feel different in the mountains because there's less oxygen in the air," says W. Ross Adey; an ardent climber and pioneer in the study of the biological effects of low-level energy fields, he conducts his research at the Veterans Administration Medical Center in Loma Linda, California. "To compensate, we have to breathe deeper, which causes us to blow off more carbon dioxide when we exhale. That allows more oxygen to get to the brain. This phenomenon, along with increases in levels of natural opiates, helps explain the 'natural highs' and unusual psychological states reported by climbers."

Just as the "rapture of the deep" has been attributed to nitrogen narcosis, and the euphoria of freezing to death, beautifully portrayed in Akira Kurosawa's film *Dreams*, to hypothermia, the profoundly altered states people have reported experiencing at very high altitudes are often ascribed to hypoxia. "When you become hypoxic, every hormone in your body changes," says Hackett. "One of the first things to occur is an increase in adrenaline, which helps explain your more wakeful state and higher energy. Somebody in a TV audience once asked me how much lack of oxygen affected my decision to go to the summit of Everest alone on my first ascent, when the rest of my team was disabled, and I'll never really know. But climbers have this saying—'The higher you get, the higher you get.' "

It is not just a coincidence that ancient religions in the world's highest mountains have made an art of breathing. Eastern meditative practices designed to achieve a natural high cultivate the outer edge of not breathing enough, which focuses the practitioner's attention by subjecting a usually automatic process to conscious control. The theory behind these exercises is that just as our psychological condition changes our breathing—anxiety, for example, makes it more rapid—breathing can change our internal state. Hackett speculates that "the feeling of relaxation and contentment people report at high altitudes and after breathing meditations might share a similar physiological basis. We know incredibly intricate details about, say, how altitude affects fine motor coordination, but not much about how it affects the mind. Whether these phenomena have a physiological or psychological basis is an unanswered question. Certainly many Buddhist monks and yogis in India and Nepal who spend a lot of time in the mountains believe there's a big relationship between environment and spirituality."

With great economy, William Blake explained part of the attraction that pulls Hackett and many others to high places: "Great things are done when men and mountains meet;/ This is not done by jostling in the street." Many of the situations climbers describe sound to the rest of us like being horribly sick in a frozen white hell, yet something keeps them going back for more. "There's a real disparity between the physical and the emotional sides of the experience, because you feel sort of chronically ill, yet emotionally you're high a lot of the time," says Hackett. "If you watch people in an altitude chamber, you can tell they're sort of in a stupor. They giggle a lot and act inappropriately euphoric. Yet they move slowly and their color isn't good—they don't look healthy. The big difference between being sick here and up there is that there, it's all self-imposed."

The testimony of both those drawn to extreme environments and those who end up in such places against their will suggests that our species still finds the occasional life-or-death crisis, so familiar to our hunting and gathering ancestors, somehow bracing. Like the prospect of hanging, the process of struggling with the forces of nature powerfully focuses the mind. Just being stuck in a car during a blizzard or caught by a sudden squall during a deep-sea fishing trip instantly eliminates several scourges of modern life. We're relieved of petty concerns and the monotony of routine, feel an immediate sense of purpose and the value of life, and perhaps even have the opportunity to be heroes.

According to Peter Suedfeld, despite the stress and strain involved, the most striking feature of human response to challenging environments is our tremendous adaptability. People involved in ordeals such as shipwrecks and plane crashes in remote areas prove to be remarkably resilient. In fact, the rates of psychological problems as well as of mortality among people who face great environmental challenges are surprisingly low—often lower than they are in normal circumstances. Sometimes extreme situations evoke the extraordinary levels of morale, cooperation, and even self-sacrifice exemplified by the popular accounts of British explorer Robert Falcon Scott's bungling but gentlemanly Antarctic expedition; on the verge of starvation with no possibility of rescue, one team member emulated an Inuit elder and walked into the frozen night so the others could have his food. Suedfeld has found that our response to an extreme setting is likely to deteriorate only after a long period of low-key frustration, confinement, and privation—the kind of chronic environmental stress faced by the Inuit each winter that is best overcome by cultural adaptation. "We've been overpsychologized to the point where we see problems and sicknesses everywhere," he says. "The fact is, most people cope pretty well most of the time."

When testing the cognitive dimensions of behavior—our ability to solve problems, think, perform, and perceive—scientists have found that particularly when survival is involved, challenging settings cause very little decline. Some slump occurs when conditions are especially extreme. For example, memory and alertness decreased among a group of Soviet aquanauts after sixteen days under water at high pressure, and, as the pilots of planes with unpressurized cabins know, above 15,000 feet, thinking gets muddled. "Near the summit of Everest, I was sure that if I jumped off, I would be able to fly," says Hackett. "I caught myself enough to realize that I was being inappropriately euphoric, and bit my cheeks and slapped my face to cause enough pain to snap me out of it. But it was a very real sensation—I truly felt that no harm could come to me. Even lower down, when cerebral function starts to deteriorate, the ability to remember, discriminate, and perform tasks quickly becomes impaired. You can't remember where the hell you put your crampons or what somebody just told you on the radio a few minutes earlier. It takes much longer to tie your boots. Then you breathe a little artificial oxygen, and all of a sudden, you're much smarter."

Research on our psychological reactions to extreme environments has similarly turned up surprisingly little in the way of deficits. The glaring exception is ability to get along with others. Most of the time, the strain a group undergoes in an extreme environment doesn't come from staring at doom, but from minor daily hassles and discomforts in what is, from the social perspective, the worst of both worlds. Despite the isolation, privacy is limited and "getting away from it all" is difficult if not impossible. During tenures in extreme settings, squabbles, cliques, and rebellions are so predictable that NASA *expects* the space-shuttle crew to get hostile with ground control. Aware of the social risks, good leaders deemphasize hierarchy when possible, stay flexible, and, like the Inuit in winter, use humor, even practical jokes, as weapons against stress. "In a tough environment like a climb, you need to be able to say, 'God, this is incredibly stressful,' " says Hackett. " 'No wonder we're feeling pushed to the max. It's minus forty and we don't have a stove and we ran out of fuel. If we don't figure out a plan, we could die.' This is an extreme example, but it has happened to me more than once. In that situation, a group can pull it together because there's a common goal and survival comes first. If people have been selected to mesh together, there generally aren't many problems. But if things have gotten a little uptight, I call a meeting and say, 'There's too much tension here. We've got to look at what's going on and talk things out so we feel better, because it's just us here, and this is starting to interfere with our work.' "

Because extreme environments are usually very isolated, the type of person who chooses to live in such places rather than just drop by on vacation tends to be a special breed. "In the back country," says Hackett approvingly, "you find more real dirt bags who can put up with a lot of stuff most people can't." Yet these stalwart souls are neither as few or peculiar as might be expected, says Rothrock. "There are a surprising number of people who live in cabins in the bush and come to town once a year. You'd think they would really be nuts, but by and large, they are not. A person can't survive out there if he's really crazy—it takes a good deal of organization and planning. Some have families, and educate their children by correspondence. I just ran into the supervisory teacher from the Yukon Flats school district, who flies out to the bush to check on the kids, and she says generally, they seem to do all right. They grow up not particularly needing a lot of social contact. They learn other ways to pass the time."

Psychiatrists suspect that one reason that the genes linked to certain psychiatric illnesses remain well represented in the population may be that some of the traits associated with them come in handy in certain settings. Anorexia nervosa, which causes mostly girls and women, many of them high achievers, to starve themselves or come close to it would seem to be one of the least productive ways to maximize the species, but Paul MacLean thinks anorexia may be a kind of adaptive behavior gone haywire: "Some researchers have observed that these women are very tough individuals—the type who could walk on forever with very little to sustain them. Migration means leaving home, and you have to be tough to do that." Because the call of the wild can ring loudest in the ears of the solitary, tough, territorial, and even paranoid, schizophrenia too may have been useful "way back in time, when one genetic strategy was being a successful loner," says Myron Hofer. "Too much schizophrenia and you're ineffective, but just a bit of it means you don't need anyone else. The invulnerable loner is one way the illness could have stayed in the population."

No matter what the peculiarities of our psychological makeup, planning and experience make life go smoother in challenging settings. "People's reactions to the Denali lab depend on how much high-altitude exposure they've had," says Hackett. "For those doing it for the first time, the isolation from culture, the lack of life, and the scale are a little alienating and overpowering. They'll say things like, 'We are so utterly cut off up here. There's no hope of rescue if we get in trouble. How are we going to get resupplied? We'll never get any mail or phone calls.' Some people actually, ah, decompensate a bit. But I've been doing it for so long that it doesn't faze me much. I get off the helicopter and, well, here I am again."

5

SUBTLE GEOPHYSICAL ENERGIES

A MONG THE small pleasures of the summer of 1991 were the frequent reports of mysterious "crop circles" that appeared in fields of grain near Glastonbury in England, an area close to Stonehenge long associated with odd goings-on. Local farmers did a brisk trade in escorting hordes of tourists to view the trampled disks of wheat, which looked something like giant elephant footprints stamped into the pastoral countryside. The media cashed in as well, filling news columns and broadcast time with explanations for the circles that ran the gamut from landing sites for spacecraft to the calling cards of whirlwinds. The accounts usually ended on a pleasingly spine-tingling First Corinthians note, reminding us that despite our know-it-all technology, we are but stewards of the mysteries of God. When at least some of the circles were finally determined to be the work of pranksters, who had trampled out the shapes with planks, the crowing of the I-told-you-so's was almost drowned out by a collective sigh of disappointment, summed up by a young woman interviewed on a radio news program: "I wish they had stayed mysterious."

Like the Glastonbury crop circles, many accounts of anomalous events are the fruit of mischievous or disturbed minds. But not all: thousands of credible witnesses over hundreds of years have seen glowing blue, white, red, or orange globes up to a foot in diameter float through the air, even through walls, before vanishing with a bang. Neither hallucinations nor UFOs, the spheres are probably so-called ball lightning. Like the conventional sort, the round kind usually appears with a thunderstorm, but because it doesn't strike between oppositely charged poles, it stretches the bounds of traditional physics. For a long time many scientists assumed that if the phenomenon existed at all, it must be optical in nature, perhaps the result of an onlooker's exposure to ordinary lightning. Pyotr Kapitsa, a Soviet Nobel prize winner in physics, was the first to suggest that ball lightning was made of plasma—electrically charged atoms of gas—briefly lit up by naturally occurring fields of electrical energy. In 1991, two Japanese scientists confirmed his suspicions, reporting in the prestigious British journal *Nature* that, using a microwavelike device, they had generated what seemed to be ball lightning from plasma. Although they still don't understand what gives the gas its spherical shape, most scientists now accept the reality of a phenomenon that had long been dismissed entirely or described as "disembodied spirits."

Technology as sophisticated as the sort that helped demystify ball lightning is accelerating research on little-understood forms of weak environmental energy and raising questions about their effects on our well-being. One innocuous testimony to this new concern hums away in more and more offices: the plug-in "ionizer," promoted as being able to approximate the bracing effects of a visit to the seaside or a mountain stream. The atmosphere of such places is thick with negative ions; these molecules of air have acquired an electrical charge by gaining electrons, usually with the help of moving water, lightning, or deposits of low-level radioactive materials in the soil. In the process of trying to shed their extra electrons to regain their neutral charge, negative ions are thought by some to create a kind of atmospheric tonic for the nervous system. In the opposite process of trying to pick up electrons, the positive ions that build up in modern man-made environments filled with artificial electricity, from airplanes to indoor malls, are thought to make us feel poorly. To defend their point of view, negative-ion enthusiasts note that we generally feel well in the kinds of natural places where they abound; certainly anyone who has ever had to flee a mall or pop aspirin in the new corporate headquarters can understand their deep reservations about the healthfulness of such places. In the laboratory, however, some studies show that positive ions worsen mood and performance and that negative ions improve them; other studies don't.

"The simple coincidence of certain kinds of ions and behavior isn't enough to establish a cause-and-effect relationship," says Louis Slesin; as the editor of the professional journals *VDT News* and *Microwave News*, he monitors worldwide research on the effects of weak energy fields. "There are lots of negative ions at the seashore and people feel good there, but we have no idea if we feel good because of the ions or because we're in a pleasant, low-stress environment. Confronting the number of things in a setting that could possibly affect us, often differently, one can only be humbled by nature. Just as there are different chemicals in the water at different places, there could be different ones in the air. There are people who have terrible allergies in the East and lose them in Arizona. In the days before we knew about such things, it must have been mystifying that a person could just go somewhere else and feel so different."

First cautioning that "it's nebulous and tenuous to relate ions to what makes you feel good," Ross Adey allows that it is possible; a leading researcher in the health effects of weak energies, he is the scientist who determined, at the request of the CIA, that the Soviet microwave bombardment of the U.S. Embassy in Moscow could interfere with the brain's ability to store and recall information. "The surfaces of cells are waving forests of electrochemical receptors for all sorts of things that affect us, like hormones and neurotransmitters. One could say that electrical charges on the ions interacting with charges from large surface areas in the lungs and pharynx could influence cells to alter the pattern of nerve impulses to the brain. There has been much speculation about this, but there is no physiological evidence that it occurs."

The experience of coediting a college textbook on environmental psychology has given Paul Bell a somewhat detached, global perspective regarding controversial issues. To say that negative ions may or may not be good for us, and that positive ones may or may not be bad strikes him as "a pretty good summary of the research. The data are very mixed. The same holds true for the behavioral effects of electromagnetic fields."

When your car radio suddenly pours out garbled static as you drive near high-tension power lines, you are moving through an electromagnetic field. These otherwise imperceptible but very real swaths of energy are created by an electrical charge in motion that has both electric and magnetic components and is capable of moving matter. Physicists and engineers traditionally maintained that the only biological effects that electromagnetic fields could produce were burns and shocks, but studies increasingly suggest that fields weaker than those of average high-voltage power transmission lines and far below what is required to produce heat can influence living things in a variety of ways. Most research centers on the potential health risks, particularly cancer, posed by artificially generated weak energy fields that emanate from industrial and military technologies as well as from ubiquitous appliances in our homes and workplaces, from electric blankets and hair dryers to computer and TV screens.

The possibility of a connection between subtle energy fields and an increased incidence of cancer first rattled the public in a major way in 1979, when epidemiologist Nancy Wertheimer found that children in her study who lived near power distribution lines, which create magnetic fields, were twice as likely to get cancer as those who did not. While some additional investigations have shown that weak fields can change blood chemistry, embryonic growth, cell division, and central nervous system function, others have not. In 1987, responding to confusion on the part of scientists as well as of the public, Congress's Office of Technology Assessment asked the department of engineering and public policy at Carnegie-Mellon University to assess all the existing—and conflicting—research linking weak energy fields to health problems. The OTA team was restricted to searching for a simple cause-effect relationship, because not even scientists convinced the fields are dangerous can say how they do their purported dirty work, or just what kind of exposure is dangerous for whom. When its survey was completed, the OTA pointed out that many experiments had found no harmful effects, but others had demonstrated changes at the cellular level that might or might not be harmful or irreversible. After stating that electromagnetic fields were a possible but not proven cause of cancer in humans, the report concluded: "The emerging evidence no longer allows one to categorically assert there are no risks. . . . But it does not provide a basis for asserting that there is a significant risk."

As this Jesuitical phrasing suggests, the question of the safety of electromagnetic fields invites a great deal more research. Lacking it, even scientists can become polarized, turning into naysayers who admit no risk at all or true believers certain the fields are highly dangerous. In the meantime, the rest of us are left to weigh as best we can the benefits and the risks of living in an increasingly technological environment webbed by invisible artificial energies.

"Everything on earth has evolved in weak natural electromagnetic fields that oscillate at very low frequencies," says Adey. "We are bathed in them from conception to death. The fields come from some very esoteric sources—things like a belt of continuous electrical thunderstorms in Central Africa and the Amazon basin, or sunspot cycles that occur every eleven years and one hundred and fifty years and are linked to increases in violent crime and stress-related deaths. We are swamping those delicate natural oscillations that have been necessary to us for so long with huge man-made electromagnetic fields whose frequencies and strengths have never existed before. And that must influence us."

When discussing the nature of our relationship to subtle energies, Robert Becker, author of *The Body Electric* and the first physician to sound the alarm about the health effects of artificial electromagnetic fields, also harks back to our evolutionary past. "When our species evolved, the earth's natural magnetic field had a frequency of one to twenty hertz. Today, America's electrical power delivery system, for example, has a sixty-hertz frequency—something that never existed on the planet until this century. The way living things work can be changed with very small electrical currents or magnetic fields, so what are we doing with all this abnormal stuff? Concern over this question has grown over the past twenty years to the point that electromagnetic fields have become *the* environmental health problem of the nineties."

Because they are rarely dangerous, less is known about the subtler energy fields produced by nature and the ways in which they too affect us, but what little scientists have learned has fascinating behavioral implications. The geomagnetic field generated by the earth's magnetic core, which sets the compass and protects us from radiation from space, is a complicated, quirky, changeable affair. Although it averages about 50,000 gamma, varying from 25,000 gamma at the equator to 70,000 gamma near the poles, some regions are riddled with geomagnetic "discontinuities," often caused by natural deposits of the conductive materials common in mountains and the ocean floor. The energy in these anomalous pockets usually measures only about twice as strong as the normal field, but if they are spread over a large area, the cumulative effect can be considerable. When approaching "magnetic deviation zones," for example, navigators and aviators know that their compass needles will be tugged from true North.

The geomagnetic field also influences the behavior of many living things. Creatures from bacteria, snails, and salamanders on up the evolutionary ladder sense and react to this mysterious force lurking beneath the thin layer of the biosphere. Migratory animals, including many birds, are particularly responsive; equipped with magnetlike organs made up of tiny collections of natural magnetite minerals, they are able to orient themselves and navigate according to the lines of the geomagnetic field. In the more responsive of these creatures, geomagnetic irregularities can create havoc. The seemingly inexplicable beachings of dolphins and whales, for example, might be caused by such aberrations, which skew their sense of direction. The geomagnetic field also seems to help regulate the biological cycles of many species. When weak magnets are attached to animals that have strong, regular patterns of behavior—say, sleep—those rhythms change in a predictable way; this alteration implies that their behavioral cycles are normally timed by cyclical changes in the strength of the geomagnetic field as the earth rotates. Like light, the earth's field may keep living things in sync with the solar system.

It has not been proved that people sense changes in or are influenced by the geomagnetic field on a regular basis, nor have scientists isolated a human magnetic organ. Unlike whales and bees, however, we might not need this special body to respond to subtle energies. The pineal gland, which regulates behavioral cycles by producing tides of neurochemicals in response to light signals, can also detect changes in electromagnetic fields. Animal research has showed that artificially generated fields, which are stronger than the earth's, can decrease the pineal gland's production of melatonin, the hormone whose fluctuations affect not only the sleep/wake cycle but also mood. Although it has not been demonstrated, it seems logical to Louis Slesin that electromagnetic fields could affect our behavior by influencing the activity of melatonin. "After all, light—what you see—is electromagnetic radiation," he says. "The idea that other types of electromagnetic energy could also affect the pineal gland makes absolute sense."

As Slesin's speculative remark suggests, research on the behavioral effects of electromagnetic fields is in its very early stages. Some studies have linked the artificial sort to increased incidences of depression and suicide, as well as to slower reactions, headaches, and lethargy, but this research has not been of the most rigorous sort. Experimentation with electromagnetic fields to treat insomnia and depression or promote relaxation, said to be popular in the former U.S.S.R., is an idea that is far from the mainstream of American psychiatry. Daniel Kripke sums up the situation this way: "I haven't seen any really good epidemiological data linking such fields to depression, and I haven't produced any significant behavioral effects by manipulating artificial electromagnetic fields in my lab. However, some serious researchers believe that the fields are changing circadian rhythms, and the data about biological cycles and the pineal gland lead me to believe that the idea that the fields could influence behavior is plausible."

Becker finds that idea far more than plausible. "Twenty years ago, despite the prevailing scientific view of the biological effects of electromagnetic fields, I knew that as an orthopedic surgeon, although I could rearrange people's parts with a hammer and chisel, I couldn't make them heal, which is a big problem after surgery. Something else did that, and it didn't appear to be chemical. Some of my colleagues and I became reasonably successful at healing bones with a little electricity, and found that an electromagnetic field that pulsed could also do the job. These things meant recognition and status, because if you can make money with something, it's got to be good. My work as a surgeon made me figure that maybe we had thrown electricity out of living things too early. Over the past decade, however, a scientific revolution has quietly been taking place. Bits and pieces have appeared in magazines and on television, but most people are still unaware of bioelectromagnetics, which is a major change in the paradigm of how science views the world. Not all, but a considerable proportion of, the global scientific community now agrees that living things are electromagnetic in nature, that their basic mechanism is a small organized flowing system of direct electrical current, and that they are tied to the earth's magnetic core, or geomagnetic field, and extract information from it. And we're just beginning to understand the pineal gland, which is so highly sensitive to the environment, and to imagine its purpose."

Becker's interest in the behavioral effects of electromagnetic fields goes back to 1963, when he collaborated on a study done at the Veterans Administration Hospital in Syracuse that showed a link between solar magnetic storms and increased admissions to psychiatric institutions. During the changes in the sun's activity that result in solar storms, massive amounts of plasma strike the earth's field, causing it to "ring" like a bell with perturbations of tiny intensity. These storms are linked to communications irregularities, unstable weather patterns, and behavioral disruptions in animals; bees alter their dances and rodents their patterns of activity. Some studies have also showed an association with increased incidences of seizures and convulsions, as well as psychiatric problems.

Entering a territory where most behavioral scientists fear to tread, Becker speculates that electromagnetic fields may even account for what are usually described as psychic phenomena. "Psychic phenomena! I hate that term—it sounds nutball," he says. "Nonetheless, the idea of messages that penetrate consciousness without coming through the senses has been around since the dawn of mankind. Most philosophies and all religions are based on the notions that there is some other reality than this and that what we perceive with the senses is not all there is." Becker hypothesizes that far from being outside the proper realm of science, extrasensory perception is part of the brain's normal functioning. In addition to our sophisticated electrochemical nervous system, he suspects that we have a more primitive communications network based on a steady flow of electrical current. Far back in evolution, he suggests, this kind of system would have allowed very simple organisms that were unable to process light stimuli to gather information from the outside world and organize their biological cycles around the rotation of the earth.

Becker's dual information-processing system would help explain the ghost-in-the-machine quandary of the so-called mind/brain problem. "The old primitive system relates us to external electromagnetic fields and controls growth, healing, and biological cycles," he says. "The much more complex nervous system provides the five senses and controls motor function. The old doesn't replace the new, but supplements it with a more basic model. Consciousness, or the mind, probably resides in the primitive system. It's interesting that yoga and meditative practices that claim to access an 'other reality' all reduce input to the nervous system, perhaps thereby increasing our awareness of the consciousness in the magnetic system—the *real* other reality. Psychic phenomena probably occur when an electromagnetic signal of some sort is received directly by the magnetic part of the brain via some as-yet-unknown mechanism."

In 1986, Becker was asked to chair a seminar at the annual meeting of the American Psychological Association. Without knowledge of one another's findings beforehand, four researchers presented papers that examined the connection between the incidence of reports of psychic phenomena and the activity of the geomagnetic field. Each study showed that when the field is perturbed by magnetic storms, which are especially frequent during the cycles of solar agitation that occur every eleven years, there are fewer reports of things like telepathy and clairvoyance, which increase when the geomagnetic field quiets. "These findings are important because they imply that unusual psychological events can be related to electromagnetic energy," says Becker. "They also suggest that the poor reproducibility of psychic phenomena could be due to a hidden variable—the geomagnetic field. And, if weak natural fields are linked to psychic phenomena, that raises the question of what effects man-made fields might have on the brain."

Often introduced to audiences as the scientist whose government grants were canceled after he talked about the dangers of electromagnetic pollution on "60 Minutes," Becker harbors some dark thoughts about the potential of this line of research. "The ability to put thoughts into people's heads without their awareness could change medicine and society, but it's also dangerous," he says. "Research is being conducted by those I'd rather not have do it. Certain government agencies have been investigating ESP for twenty years, but the results have not been made public. Science is on the verge of important discoveries, and we can't leave them up to unpleasant agencies that could misuse them."

At the NIMH, the subject of the effects of electromagnetic fields on behavior generally elicits the kind of response that questions about the healing power of crystals or pyramids might warrant. "It makes sense that energy fields coming from computer screens and other appliances could affect the nervous system, which is electrochemical in nature," says light expert Norman Rosenthal, the only researcher there willing to speculate about the issue. "But when you're sitting in front of a computer, you're subject to many other environmental factors, from poor posture to eyestrain, that could also affect mood and well-being. The study of electromagnetic fields, sunspots, ions, and all those things is legitimate and sensible, but there hasn't been enough research that has isolated their effects, which is very hard to do. It's not that such energies don't affect us, but that we don't know whether they do or not. The definitive work is yet to be done."

While discussing some of the peculiarities of places rich in electromagnetic anomalies, Becker recounts the story of a hunting party of medical professionals he knows who fled from a remote valley in New Mexico when they suddenly experienced profound distortions in color perception—for example, cacti turned red. Hypothesizing about what people in simpler times would understandably have considered a supernatural experience, Becker suggests that an anomaly in a mountainous region with a lot of magnetic materials in the ground suddenly changed the geomagnetic field in a small area; upon entering it, the activity of the hunters' pineal glands was altered, and perhaps that of an electromagnetic organ, which caused neurological changes that altered perception. "This kind of thing is hard to study, because electromagnetic distortions vary constantly and may only affect an area of a hundred feet," he says. "But we know animal migration can be affected by these discontinuities, and all living things that have cyclical functions might be sensitive to them. The point is that there's a relation between the activity of the brain and the electromagnetic field at a given location that makes us creatures of the geomagnetic field in a deep subconscious as well as biological sense."

6

■

SACRED PLACES

D URING 1968 and 1969, hundreds of thousands of people reported seeing the Virgin Mary and other celestial beings over a Coptic Orthodox church in Zeitoun, Egypt, not far from Cairo. Accounts described different sorts of visions, including "doves"— small, moving, short-lived lights—longer-lasting corona-type luminous displays, and detailed apparitions. The events took on an extra *frisson* because they could not be entirely dismissed as the products of overheated imaginations: photographs taken at the site actually showed glowing blobs of light. When Michael Persinger, a professor of psychology and neuroscience at Laurentian University in Sudbury, Ontario, examined seismological records, he found that the Zeitoun visions began a year before an unprecedented increase, by a factor of ten, in seismic activity 400 kilometers southeast of Cairo.

After reviewing the circumstances of 6,000 strange events of the sort usually labeled either supernatural or fraudulent—fish or frogs "raining" from the sky, UFOs, haunted houses, poltergeists—Persinger has found that many lend themselves to natural hypotheses, if not explanations. Although he primarily studies brain function, Persinger has a background in geophysics as well, and this combination of interests inclines him to think about how the earth's processes affect the nervous system. Starting with one of the bizarre phenomena that most scientists steer clear of, such as the lights at Zeitoun, he "works backward to see how the brain is put together. We're conditioned to think that fish stay in water, that rocks don't pop out of the ground, and that odd lights don't suddenly appear, so when we're confronted with anomalous data, we try to cram them into best-fit scenarios. Both religion and science provide structured ways to do that. Until the development of plasma physics, we had no scientific way to think about ball lightning, but that doesn't mean it was caused by demons."

After grappling with the dilemma of how to study extraórdinary events that are by nature infrequent, short-lived, and highly localized, Persinger has learned to look not at isolated incidents but at the patterns into which they often fit. After plotting data concerning location, time, and simultaneous geophysical activity, he finds that the kinds of weird phenomena associated with particular places are often linked with unusual perceptual, chemical, or energy-field stimuli, either of a "chronic" or "acute" nature. "You have to wonder how many historical events have been shaped by religious interpretations of freak phenomena, as occurred at Zeitoun," says Persinger. "When Constantine saw what he perceived as a giant cross in the sky, the Roman legions followed him to victory, which made Christianity the state religion of the empire. For that matter, would Christ have assumed such a place in history if he hadn't died during an earthquake, when 'there was a darkness over all the earth . . . And the sun was darkened, and the veil of the temple was rent in the midst'?"

One evening, an experienced geographer who works with Persinger was sitting in her car in remote country. She noticed that everything had grown very quiet, and that an ozone-scented breeze had begun to stir. Then, a light shone in her rearview mirror, followed by a nebulous glow that permeated the car. Although she had tingling sensations and knew she should move, she didn't because she felt so euphoric. Falling over sideways, she looked out the window and experienced a hallucination of a dog suddenly appearing from thin air. About fifteen minutes later, when she came to, she found that her car's alternator had been burned out, but other than feeling queasy for a few days, she was all right.

Persinger numbers the geographer's strange tale among the bizarre one-time experiences in natural settings that he classes as acute, and often attributes to geophysical anomalies. "Profound perceptual changes, such as hallucinations, can result from the induction of substantial direct current into the body," he says. "That's why many people who've had such experiences report that they woke up after being unconscious—they'd been knocked out," he says. "If the current, perhaps generated by tectonic strain deep within the earth, is too intense, the person dies of a heart attack or seizure, and is reported as having had one of the normal kind."

Employing twentieth-century technology and the help of more than 150 research subjects, Persinger has orchestrated laboratory versions of the heightened experiences that have throughout history drawn people to religion, art, and drugs, as well as special places. To uncover the prosaic neurological wiring behind sudden bursts of illumination and emotion, he puts his volunteers in a novel environment, then delivers to their brains a pulsed magnetic field of the same intensity as that of a commercially available relaxation device. Although he is primarily interested in studying brain function, on his clinical rounds at the university's cancer clinic Persinger sees a potential practical application for artificially induced "mystical experiences" of the kind his subjects often report. "People have become more secular, but they still have a lot of anxiety about death," he says. "One dimension of our work is to find ways to let the dying have an experience that will reduce that suffering. So what if it's synthesized?"

When it comes to altering consciousness, Persinger has found that "set and setting"—state of mind and physical milieu—are as crucial as electrical stimulation to his subjects' experience. He is hardly the first to have observed the connection between extraordinary environments and perceptions. The psychedelic light shows that accompanied rock concerts in the 1960s and 1970s are modern examples of settings designed to "blow the mind," but they have far earlier precedents. "At the time Gothic cathedrals were designed, most people lived in dark huts, so just walking into a space vastly larger than what they were habituated to, lit by stained-glass windows, was literally awe-inspiring," says Persinger. "Today, we're not as impressed by big buildings, so we have to go to very large mountains to experience that 'diminutive effect.' Nor had medieval people heard anything like the acoustic effects of two choirs responding to each other from opposite ends of the church, which would have flooded their senses, shifting transmitter levels and releasing natural opiates. The more vulnerable might have been driven into ecstatic states, the extreme form of which is the seizure, which was commoner then because of bad nutrition."

Anyone who has ever crawled off a plane after a long flight only to perk up as if by magic at the sight of the Eiffel Tower or the Grand Tetons knows something about the power of a new environment. "Our experiments are conducted in an acoustic chamber that's tinted with red light and filled with eerie music of different sorts," says Persinger. "This setting produces a strong feeling of novelty, which jacks up the subject's adrenaline. We know that if someone is injected with a small amount of it and then put in a setting where everyone is crying, he'll get upset, too. If he's in an aggressive setting, he'll become irritable. Similarly, we've found that for a person aroused by a novel setting, tapes of Gregorian chant are likely to produce a religious experience, while the *Close Encounters* theme inspires a UFO-type experience. In short, in an aroused person, the cognitive aspects of a situation determine his emotional response."

Differences in our nervous systems, which are electrochemical in nature, incline some of us to be more emotionally responsive than others. Studies of epileptics have linked the greater electrical volatility of their brains to the unusual thoughts, sensations, and emotions that often precede or accompany seizures—perceptual alterations that closely resemble those experienced by Persinger's subjects. He suspects that to a far lesser degree, others' brains are electrically labile, too. "The people I'm talking about tend to be energetic, creative, suggestible, intuitive, and to feel 'guided,' " he says. "They're really normal people, except for slight hypomanic spikes that suggest an excitable brain chemistry. Ultimately, all our experiences are tied to our transmitter balance. Unusual neurochemistry, unusual experiences."

Although his volunteers' accounts indicate that what they experience during Persinger's mystical experiments depends as much on set and setting as on electromagnetism, animal research has shown that more powerful electrical stimulation of particular brain structures instantly produces certain behavior, from displays of aggression to feeding to sexuality. The Spanish neuroscientist José Delgado has even stopped a charging fighting bull in its tracks by activating a remote-controlled electrode in its brain. Very little of this kind of invasive investigation has been done with people, but in a series of very controversial experiments, Robert Heath, a neurologist at the Tulane Medical School, elicited ecstatic feelings and sexual climax by electrically stimulating the brains of incurably ill subjects; he found that the experience of orgasm corresponds with a massive electrical discharge in the septal region of the limbic system, the brain's emotional center.

The electromagnetic signal Persinger's subjects receive particularly activates two of the most electrically sensitive structures of the limbic system: the amygdala, involved with assigning meaning to sensory input, and the hippocampus, involved with memory. Simultaneously shaking up the neural foundations of memory and meaningfulness can suddenly release a flood of images from the past that are automatically imbued with a tremendous sense of reality and importance. Persinger offers the so-called bereavement hallucination, one of the most commonly reported mystical experiences, as an example. He has found that these visions of the deceased usually occur soon after the death and on days of significant geomagnetic activity, when episodes of epilepsy and psychiatric hospital admissions increase as well. This coincidence of the internal arousal of the bereaved and some external stimulation is another variation on the theme that brings about unusual experiences in a laboratory or the wilds.

"Electrical stimulation just increases the intensity of what a person anticipating an unusual event in a novel setting would experience anyway," says Persinger. "He would be inclined to report feeling a bit odd or dizzy, the presence of someone or something, tingling sensations, fear, happiness, dreamlike images. With some stimulation, those phenomena grow stronger, and some new ones show up, too. Some subjects might think they're getting ideas from outside their own minds, experience odd tastes, entertain thoughts from childhood, or have the sensation of being detached from the body."

The symptoms Persinger describes are just the sort often reported by those who have had an acute encounter with an "unidentified flying object." At the turn of the century, what we call UFOs were referred to as "odd airships," before that "odd luminosities," and before that, angels or demons, depending on the context. After reviewing more than 1,200 reports, 99 percent of which concern balls of light that move about, Persinger believes that most UFOs can be explained, and even predicted, by solar and seismic variables. "Between August 2 and 7 in 1972, for example, massive sunspot activity shocked the earth hard enough to knock it off its orbit," he says. "On August 10, multicolored fireballs, probably consisting of plasma, were reported over both countries. On August 19, a major UFO flap began, with many people reporting luminous objects, football-shaped spacecraft, and the like."

The most flamboyant accounts of UFO experiences are those given by a growing number of people who describe themselves as "space abductees." The kind of event they report goes something like this: A person driving along in a remote area feels a force push his car off the road. When he gets out, he hears weird noises, or sees rocks leaping from the ground or an odd rain falling from a clear sky. His car engine suddenly stops, and a UFO appears. As he draws closer to it, he passes out. Later on, he feels unwell and perhaps somehow violated, but can remember little or nothing of why that might be.

Persinger offers this hypothesis for the space-abductee experience: "If an electromagnetic field produced by tectonic strain is intense enough, rocks can move about and water vapor can fall from low-level ionization of the nearby air. Prefracture strain can induce rumblings, groans, and pops. A force generated by accumulated tectonic strain could pull a car off the road. If the air is ionized, as happens in ball lightning, a luminous blob that could be interpreted as a UFO might appear. If the intensity of the electromagnetic discharge wasn't dissipated by ionization, it could deliver a current briefly to the earth, causing the observer to see a luminous column slowly descend and land, à la *Close Encounters of the Third Kind*. (Devil's Tower, the Wyoming site of the UFO landing in the film, was a sacred mountain of the Lakota Sioux, who conducted rituals there at the summer solstice.) The commonly reported phenomena of blacking out upon drawing close to the display and subsequent amnesia suggest an assault on the brain's electrical system, which could also cause the car to fail. When the electrical field finally discharges through intense ionization, the 'UFO' disappears long before it can be documented."

As those who have seen *Close Encounters* recall, some people who have had a full-tilt UFO experience, such as the space abductees, feel it has changed them and the way they look at life. Persinger suspects that this kind of psychological shift probably indicates that the electrical effects of tectonic strain were focused on the onlooker's brain. By way of illustration, he suggests an explanation of the grisliest part of the abductee stories: the narrators' claims of having been kidnapped by sinister aliens and subjected to strange experiments, often of a reproductive nature. "Powerful stimulation of particular regions of the brain can evoke the feeling of a presence, disorientation, and perceptual irregularities," he says. "It could also activate images stored in the subject's memory, including nightmares and monsters, that are normally suppressed. When the parts of the brain that certify an event as authentic and meaningful have also been stimulated, the subject's psychological experience—perhaps a feeling of being in touch with a supernatural power or another world—would seem very real and emotionally charged." Regarding the physical symptoms sometimes reported, it's possible that ionizing radiation could cause skin problems, nausea, fatigue, depression, and endocrine or gonadal problems. What the subject recalls later may reflect the body parts affected by the electricity. For example, says Persinger, proprioceptive changes in gonadal tissue may show up as the commonly reported recollection that 'spacemen did tests on my genitals.' All these unusual perceptions are apt to be interpreted anthropomorphically as a feeling either of being under surveillance or attacked, or specially protected."

Religious shrines throughout the world attest to the fact that when a person experiences extraordinary perceptual change—say, a heavenly being appears out of nowhere—along with a sudden revelation, he and even his society may consider the place where these things occurred special. Albert Einstein may or may not have felt that way about the setting in which $E = mc^2$ suddenly "came" to him, but that kind of instant insight, in which the long-snarled strands of some dilemma suddenly untangle, is a good example of what Persinger calls the "2 A.M. wow." "That feeling you get when it all comes together results from normally occurring excitation in the amygdala, one of the brain's seats of memory," he says. "Imagine that you're out walking through the boonies in some special place that has a magical reputation. Suddenly, a natural force, perhaps set off by a geomagnetic storm, kicks in and gives you a much more powerful version of that sudden revelation. It's no wonder that most people rate these 'sacred place' events as very meaningful and positive. Even our volunteers in the lab, who get a much weaker zap than they'd get outdoors, want more of the artificially produced version."

While Persinger has catalogued a surprising number of singular experiences of the sort reported by his geographer colleague and the space abductees, many more reports concern milder events of a chronic nature. Not surprisingly, the locales in which they repeatedly occur often acquire reputations as sacred or tabu places. Since history's first epic poem recorded the visit of the Sumerian hero Gilgamesh to a special grove of cedars, certain natural spots scattered round the world—Ayers Rock, Mount Fuji, Canyon de Chelly, the springs at Lourdes, the Ganges River, and hundreds of others—have drawn people seeking insight, inspiration, healing, or proximity to the divine. Often, the same places have been revered by very different societies. Jews, Muslims, and Christians alike venerate Mount Sinai; the California hot springs that incubated many of the spiritual and cultural changes of the 1960s were once sacred to the Esalen Indians; and many of Europe's cathedrals were deliberately built over pagan springs and ritual sites. Powerfully augmented by the pilgrim's expectations, it seems that the special physical properties of what the Bible calls "high places" have the capacity to promote physical and psychological change.

Recently, hordes of upscale spiritual seekers, drawn by claims that the former sacred mountains of the Incas foster mysticism, have been flocking to high places in the Chilean Andes. The obliging peaks even periodically blaze with weird glows and flashing lights, accompanied by popping, sizzling sound effects. To the swelling community of New Agers, this mysterious son et lumière is tangible proof that they are at a hot spot of spiritual magnetism. Recalling that the "Andes lights" were first reported in a 1912 issue of Scientific American and more recently by Gemini astronauts, scientists suspect electromagnetism instead. Anomalous energy fields are generated not only by the conductive sediment in mountains but also by that in the large basins considered holy places by Native Americans, including the Uinta in Utah, the Tucumcari in New Mexico, and the Black Hills in the Dakotas. "We're electrical beings living in a magnetic environment," says Louis Slesin. "Because we're finely tuned to subtle energy fields, when they vary, as they would on top of a mountain, we change biologically and psychologically too."

Of the chronic phenomena associated with sacred or tabu places, strange lights are perhaps the most frequently reported. In these areas, astonished visitors may encounter luminous orbs that barely warrant a raised eyebrow from residents accustomed to "ghost lights." One to four feet wide, the glowing balls tend to show up repeatedly in the same place, retreating when pursued. The Comanches attributed the West Texas version filmed by a CBS news crew to the spirit of a long departed chief; New Jerseyites claim the "Hookerman ghost light" is the specter of a railroad man killed along the tracks where the glow appears. Persinger, however, believes that ghost lights are simply the luminous electrical manifestations of geomagnetic anomalies or the focal points of tectonic strain. "When an electrical discharge is concentrated in a spot that allows the maximum field and ionization potential—often at the tops of hills and buildings, near power lines, or in swampy overgrown areas where decomposition releases combustible gases—strange lights and power failures can occur," he says. "These magnetic-field zones can also cause odd psychological reactions. There are written accounts of people who have stepped into such an area, felt fearful, stepped back out, and felt all right again."

Beneath the earth's surface seethe massive geophysical forces that constantly interact with and change it. This activity can produce acute events of the kind sometimes described as "acts of God." In 1920, in Lincolnshire, England, a brook "jumped" twenty feet, killing fifty people; in 1968, a fourteen-foot-wide, fifty-foot-deep hole suddenly appeared in a backyard in San Diego; in 1973, several tons of rocks, dubbed "earth cookies," popped from the ground in Elk Hill, Oklahoma. During the far more violent shifts in the planet's crust responsible for earthquakes, tremendous seismic pressure pushing on rock crystals produces electrical fields across large areas on the ground and in the atmosphere. These fields can measure several thousand volts per meter, an intensity capable of creating the well-documented phenomenon of "earthquake lightning." According to Persinger, many "unidentified flying objects" are, like the celestial beings of Zeitoun, essentially earthquake lights for very small earthquakes. "The anomalous luminous displays strongly suggest tectonic strain in the locale," he says. "The resulting electromagnetic fields can directly stimulate some observers' brains, provoking psychological phenomena reinforced by their own personal histories. That's why at Zeitoun, one person 'saw' the Virgin, while another saw a dove." When geophysical perturbation continues over time, it can add to the reputation of a sacred place. Between 1972 and 1979 alone, for example, there were 82 reports of luminous phenomena near Toppenish Ridge in south-central Washington; the ridge, part of the Yakima Indian Reservation, lies within the Yakima fold belt and is still undergoing compressional deformation.

Chronic environmental weirdness isn't limited to natural settings. Just about the time when the Glastonbury crop circles were making headlines, a judge ruled that a woman who owned a house in the New York suburb of Nyack was remiss in not informing the resentful party to whom she sold it that the place had a history of being haunted. After investigating 207 cases of the architectural version of ghost-light zones, Persinger thinks the moving objects, fires, and odd noises associated with haunted houses and poltergeists are prompted by energy released by tectonic strains and geomagnetic activity, which affects various materials, including the human body, according to their properties, such as conductivity. "Some haunted houses have had a hundred owners over thirty or forty years," he says. "They often report trouble with lights blowing—they just can't keep incandescent lights in the house. The unusual emotional and perceptual experiences they report resemble those described by people whose brains have been stimulated during surgery or in attacks of nonconvulsive epilepsy. In both instances, subjects report strong smells, loud noises, depersonalization, and dreamlike visions. Depending on the focus and intensity of the stimulation delivered to the brain, the experience could be pleasant or terrifying. This combination of real external physical events and unreal ones from the stimulated brain produces the confusing mixture of the rational and bizarre typical of accounts of haunted houses."

Although few of us experience cacti turning red, animals appearing from nowhere, hauntings, vivisecting aliens, or Einsteinian revelations on mountaintops, our everyday perceptions also spring from a collaboration between our brains and environments. This feedback process begins at the earliest stages of life, in the surprisingly complex world of the womb.

INSIDE OUT

.

PART II

.

7

.

PERSON AS
PLACE

G REGORY, A preemie finishing what should be his gestation in a
neonatal intensive care unit of New York's Columbia–Pres-
byterian Medical Center, is one of the youngest people on the planet,
but he seems like one of the oldest—a trembly, wizened grandpa
with a puckered, sunken face and big ears. At his birth two months
before, twelve weeks ahead of schedule, Gregory was a foot long and
weighed a pound and a half. Five lines thread through the portholes
of his aptly named isolette to his body: oxygen tubes, a heart monitor,
IVs for feeding and medicines. To keep them secure and untangled,
one of his arms is taped to a board and his feet are restrained by
bandages clipped with big safety pins to the sheet.

The stainless steel atmosphere of the neonatal intensive care unit,
or NICU, is necessarily designed for lifesaving medical procedures.
During the periods of wakefulness he is not meant to handle yet—
a fetus of his maturity sleeps 95 to 99 percent of the time—Gregory
responds to his grueling institutional environment. The mere click
of his isolette's door sets him twitching all over. In the round-the-
clock glow of the fluorescent lights, his eyes slit open from time to

time and immediately wince shut. When his nurse wipes his nose with the corner of a small gauze pad and calls him pumpkin, Gregory frowns and flinches. In this sterile setting, what comes across most poignantly is the deeply sentient nature of this tiny specimen of Homo sapiens, still a long way away from the life of the mind that distracts his elders from the life of the body.

Watching him, the assumption that preemies are simply younger, smaller, sicklier newborns soon evaporates. Each moment that passes strengthens the uncomfortable perception that very much like a fish out of water, Gregory is struggling to live in a world for which he is unsuited. Having been born, he is called a baby. However, says Evelyn Thoman, the director of the infant studies laboratory at the University of Connecticut at Storrs, "the fetus and the preemie are very different sorts of beings that haven't been designed by evolution to survive outside the womb. There's certainly no other animal like the preemie. It's really a new species, a totally unique twentieth-century organism."

In the effort to engineer a habitat for this special modern organism, scientists are revolutionizing our understanding of the secret world of the womb and the fascinating creatures we were when we lived there. Until very recently, the uterus was regarded as a kind of static puddle in which the passive fetus floats, largely oblivious to what is going on around it. Delicate researches show that far from waiting inertly for birth to raise the curtain on environment, from its first cellular murmurs the fetus is profoundly attuned to the uterine world and, gradually, to the one beyond. From these early days onward, its development and behavior will continue to depend as much on the place in which genes are expressed as on the genes themselves.

As the hands-on world of the child recedes, we tend to forget that the body isn't just a hat rack for the mind, but the crucible of development and the creator, monitor, and synthesizer of all our experience. We come to rely heavily on vision and hearing, the senses that service the modern workplace, yet we evolved with head-to-toe sensory organs, just as important as eyes and ears, that have enabled our species to adapt to a huge range of environments, from the tropical forest to the poles. One of the most eloquent illustrations of how the senses and the environment collaborate to shape development and behavior are the records of feral children. Abandoned early in life and left to shift for themselves as creatures in the wild, they grow up apt to be mute, almost insensitive to pain, unable to laugh or smile, and to walk on all fours; however, they can track people and things with a finely developed sense of smell, a capacity the rest of us rarely cultivate. If an observer from another planet were to base his impression of a human being on a feral child, we wouldn't recognize his description. Yet the contrasts between us and these weird creatures are not genetic, but spring from two very different early experiences with the environment.

During most of the time since the invention of the incubator in the early part of the century, it would not have been possible to observe Gregory in the environment of the NICU, perhaps even for his parents. Despite the fact that preemies have been popular exhibits at world's fairs, in hospitals they were generally protected from outside stimulation as if it were a disease. Then came the 1960s and the dawn of a powerful idea that affected thinking in science as well as culture: an individual and its environment were best understood not as separate entities but as a dynamic feedback system. This fertile concept popularized phenomena as various as ecology, the "open classroom," and "birthing rooms" with mandalas on the walls and whale songs on tape. In that ebullient era of More Is More, scientists working with preemies grew concerned over some seminal animal studies that showed that sensory deprivation in early life stunts behavioral development.

In the 1950s, University of Wisconsin psychologist Harry Harlow separated infant monkeys from their mothers and raised them in various barren environments. The most deprived developed severe behavioral abnormalities and shrank from social contact. However, monkeys who had a padded feeding apparatus, or "mother," in their bleak quarters fared better than those provided only with a wire model. In later experiments, researchers found that monkeys given a stuffed feeder that also rocked did best of all, growing up free of many of the aberrations the others suffered. These sobering insights into the impact of early environmental stimulation caused scientists to wonder if the developmental troubles preemies experience were aggravated by long tenures in sterile NICUs.

Determined not to foster a kind of high-tech Harlow monkey or feral child in the NICU, researchers trying to help preemies started what Thoman calls the "What else can we do for them?" movement. "First we let the parents in," she says. "When the babies did no worse, we did more and more, as if we could replace all the different kinds of input, both fetal and neonatal, that they were deprived of." These early grab-bag stimulation programs, in which preemies were held, talked to, and stroked, often all at the same time, seemed to make many infants and certainly their caretakers feel better. However, no one was sure exactly what kinds of stimulation each baby was getting, or which were helpful or even harmful.

In an effort to find out, Gerald Turkewitz, a professor of psychology at Hunter College, and some colleagues decided to monitor everything that went on in an NICU every fifteen minutes for seventy-two hours straight. Their findings surprised them. It turned out that the preemies were handled about 10 percent of the time, which is about the same amount of touching that babies get at home. But the lighting in the room never went below the level required for reading a newspaper, and certainly did not vary by day and night as it would at home—changes vital to our biological and behavioral rhythms. And 90 percent of the time, the nursery noise level was the same as that on a busy street corner. "The loudest rock music I ever heard was at two A.M., when the nurses were trying to stay awake," says Turkewitz. "There are differences in what we and babies perceive, but certainly the notion that preemies get insufficient stimulation is just all wrong. In fact, they mostly get too much."

To determine the consequences of this sensory barrage, Turkewitz made a few modest changes in some NICUs—things like dimming the lights at night and shutting doors to cut noise. Upon their discharge from the hospital, evaluations showed that the babies from the modified nurseries were better at organizing behavorial states, such as sleep and wakefulness, he says. "What was particularly striking, however, was that responsiveness to sound was depressed in all the babies from noisy NICUs, but much less so in those from the quieter nurseries."

There was nothing wrong with the ears of the unresponsive babies. Their apparent insensitivity was caused by an environmental obstacle that had prevented them from making an important connection between the senses of hearing and sight. At home, when a baby hears a sound, he turns his eyes toward the noise and usually sees what made it. This audiovisual sequence helps him start to make sense of the world: "When I hear that click, the door opens, and Mom appears." But in one NICU Turkewitz studied, many sounds the preemies heard came through doors that opened onto a busy corridor. A baby would hear a noise and turn toward it, but get no visual reward for his trouble. Before long, he learned not to bother looking for connections between sounds and sights. An important foundation of his development had been weakened by the place in which the process had started to unfold.

"The problem isn't that such a baby is understimulated, but that the world is subjecting him to the wrong kind of stimuli, often before he can handle them," says Turkewitz. "Because evolution scaffolds one stage of development on another, inappropriate stimulation of one sense can even cause problems with another. For example, rats whose eyes are opened too early end up with an impaired sense of smell. One of the advantages of the uterus as a place in which to develop is that it offers limited sensory input. That means the baby can slowly build up a developmental framework and keep tacking on to it."

Less concerned about the perils of understimulation, researchers working with preemies now worry that trying to develop in an NICU is like trying to live in a busy airport or subway station. The problem isn't that these places offer too little, but, in a peculiarly modern way, too much. Preemies are not only born before they can handle the normal stimulation of a home and family, but also spend weeks or months in an artificial milieu that would badly stress a robust adult. For both these reasons, "preemies don't seem able to get in sync with the environment very well," says William Fifer; a developmental psychologist at the New York State Psychiatric Institute, he studies babies' neurobehavioral development ("not quite the mind, but almost"). "As a result, they're much more irritable. Their cries are very shrill—it's no coincidence that they get abused more. Later on, they're likelier to have more developmental and learning problems. If their sensory environment could be manipulated in the right way, they could organize themselves better."

So could we all. It's striking how often the observations of scientists trying to design the perfect place for preemies also apply to the rest of us in our jumped-up, fast-forward world. "Like any group of people, preemies are an extremely heterogeneous population with very various histories and constantly changing needs," says Thoman. "These days we've become more interested in protecting them from overstimulation again. There have been lots of these ideological swings, and they're all right and all wrong. We overstimulate them and we understimulate them, and sometimes we get it right. The question really is 'What is the best kind of stimulation, for whom, and when?' "

The resounding theme of our relationship to the environment before birth applies throughout our whole lives, from the cradle to the schoolroom, the home to the workplace: our well-being depends on the delicate business of getting just the right amount of stimulation from our surroundings at the right time. Yet this scientific reaffirmation of the old Greek principle of moderation runs counter to cherished modern notions of progress. "There's a common belief that if some stimulation is good, then more must be even better," says Anneliese Korner, a professor of psychiatry and preemie researcher at Stanford University School of Medicine. "There's also an assumption that by trying to accelerate development, you can produce a more intelligent or proficient person. There is no evidence at all that either of these things is true."

The unique feature of our first place is, of course, that it is also a person. For nine months, we dwell in the flesh and fluids of a fellow sensate being who is literally our whole world. This combination person/place handles everything for us, even the climate; the placenta distributes extra heat from the fetus, who is usually a degree warmer, to the mother. "We still don't know what it's really like in the womb," says Fifer. "But we do know that an externalized fetus, a twenty-five-week-old preemie, does much better with its soothing, muffled sounds heard contingently with movement, its steady temperature, its gentle changes in light, and its constant gravity. One way to think about what life inside a person is like is to observe how different even one aspect of it, say, movement, is for a preemie. He no longer rides around in his mother's belly, rocked and rolled day and night. When he moves on his own, with very little muscle strength, he must gesture in the air, a very different experience from sailing an arm or leg through amniotic fluid, hitting the uterine wall, and getting feedback."

Living inside someone else has its advantages, but it has its problems too, as ubiquitous warnings that pregnant women ought not to use drugs make plain. The placenta delivers oxygen and nutrition to the fetus and returns carbon dioxide and wastes to the mother for elimination, but the drugs and toxins the mother eats, drinks, breathes, or otherwise takes in travel the same route. Because of the damage that drugs can do, some maintain that women who indulge should be imprisoned during pregnancy—a political position that frankly equates mother with incubator. New inquiries into the consequences of even the most health-conscious pregnant woman's way of life are destined to raise even more difficult questions.

It's a rare expectant mother who is not advised by a well-meaning observer or two to skip funerals, listen to classical music, avoid confrontations with the handicapped, or in some such way modify her psychological experiences for the good of the unborn. Such admonishments have been dismissed as medieval superstition by modern medicine. Recent animal research, however, shows that at least one sort of psychological experience—stress—can have lasting effects on a pregnant female's offspring. "One of psychiatry's basic principles is that early experience is crucial to character and behavior," says Thomas Insel, a research psychiatrist at the NIMH who studies the biological substrates of parental behavior. "It's well established that the loss of a mother or father is linked to long-term depression, even suicide, and that the development of speech, vision, and sociability depend on having certain experiences in infancy at the right times. What we're learning is that the effects of these events aren't just psychological, but neurobiological as well."

During certain stages in the development of the fetal brain, a pregnant female rat's brief confinement in a stressful environment—one that is too crowded or too bright—somehow shifts her neurochemistry, which in turn alters the maturation of her offspring's nervous system later in life; as a result, males, for example, show female sexual behavior. The brains of such rat pups also have fewer than the normal number of receptors for opiates and more for benzodiazepines, or natural tranquilizers; it is as if their neurophysiology reflects the low-pleasure, high-stress experiences undergone by their mothers. Supporting studies show that when pregnant females are artificially aroused with a dose of norepinephrine, a stress hormone, their pups grow up likelier to be timid in new situations. "Neurochemical and behavorial changes in offspring can be induced by maternal experiences, just as they can by drugs," says Insel. "Stress in particular seems to shake up the same systems affected by harmful substances."

Even though it is in its early days, research on the effects of prenatal maternal stress on babies has already been politicized enough to put off many scientists. One study done at UCLA and the State University of New York at Stony Brook showed that women who experience distress and anxiety during pregnancy are likelier to deliver premature and smaller-than-average babies. Because the implications of such findings are so upsetting to so many, scientists probing the chemical environment of the womb are the first to urge caution in extrapolating advice for pregnant women from animal data and preliminary human studies. "A mother's hormonal and emotional state probably has many influences on the fetus," says Myron Hofer, who studies the mother-infant bond in animals. "But considering the number of normal children whose mothers went through ghastly pregnancies, there's clearly lots of buffering in the womb as well."

Considering that research turns up ever more ways that even the most diligent woman can jeopardize her fetus, it seems only fair to let the record show that the fetus doesn't always make life easy for its hostess, either. "In their last trimester, women carrying boys report more discomfort and irritability than those pregnant with girls," says Thoman. "There's a complementary finding that primates carrying females get beat up more often than those bearing males because they're less aggressive, probably also for hormonal reasons. From the beginning, there's an interaction between child and mother, creature and environment."

On first visiting Gregory's NICU, adults must be a bit disappointed that no Disney murals and splashes of primary colors enliven the walls. The only visual cheer radiates from a few isolettes decked with bright paraphernalia imported by wistful parents. "It makes the families feel better to bring snapshots and stuffed animals, but the babies don't seem too interested," says Mary Banfield, director of nursing. Then she adds gently, "although sometimes they do like to be stroked with the fur."

Because we're so visually oriented, it's hard for us to accept that new babies aren't. Even full-termers have poor sight beyond the distance of a mother's face during nursing. Certainly visual stimuli are subdued in the world from which they've recently emerged. "We know the fetus can see, because if we shine a bright light through the mother's abdominal wall, it will blink," says Fifer. "It probably can't make out shapes, but perhaps it can see the uterine walls and its own limbs, blurred as they would be under water. And it can pick up the changes in day and night illumination that are so important to our patterns of activity and mood."

Anyone who has watched a spy movie with a brainwashing scene or been stuck for the night in an air, train, or bus terminal knows that round-the-clock illumination is exhausting and disorienting. Aware of the havoc constant light plays on circadian rhythms, nurses in the NICU shield some babies from the twenty-four-hour glow by tossing blankets over the tops of their isolettes. In the nursery where older toddlers recuperate, the blinds are pulled to ease naptime. Providing these gentle changes in illumination is far more important than hanging up bright decorations, which might even be harmful, says Thoman. "A while ago, someone observed that six-month-old babies like bold, bull's-eye-type graphics, so we put them in with preemies. What we didn't know was that unlike older babies, these little ones have an obligatory eye response that compels them to stare at such images, which ends up stressing them."

While the womb serves as a shield that allows the gradual development of sight, it functions quite differently where hearing is concerned. At one time or another, we've all reclined in the bath, ears immersed, and listened to the household noises around us. Perhaps part of the charm of this simple experiment is that it captures something of our first experience of the world: its sounds. The womb is a surprisingly noisy place. Most of its volume level of 90 to 95 decibels is produced by the rush of blood through the uterus, the maternal and fetal heartbeats, and intestinal gurglings. Along with all this internal racket, which on tape strikes adult ears as a loud, throbbing audio version of Chinese water torture, the womb also admits a good bit of what is going on outside, especially speech, and most particularly, the expectant mother's.

The human voice is an attractive stimulus because unlike, say, the hum of an air conditioner, it is complex. A mother's speech, however, has a special fascination for her fetus. Not only is the sound both air- and fluid-borne, but it also provides a unique multisensory experience: every time a mother speaks, her fetus gets a massage. Considering all this excitement, it is not surprising that the newborn recognizes his mother's voice before her face or smell. A day-old baby wearing headphones that play his mother's voice quickly learns that the sound will stay on as long as he sucks a pacifier-monitor, and he sucks more for that voice than for others. "At first the baby prefers tapes of a filtered version of maternal speech, the way it sounded in the womb," says Fifer. "By the end of the week, he prefers her real voice."

Even in the NICU, a baby often recognizes his mother's and even father's speech, and will calm down if a parent talks to him during an upsetting medical procedure. Because of the soothing power of the maternal voice, says Fifer, "tapes could be played to help these babies feel better, but first we have to know more about when to do so and for how long." He suspects that maternal speech also plays a role in the infant's early learning, perhaps by helping to teach him to tune in to some cues and turn off to others. Despite the importance of the fetal and neonatal acoustic environments, however, one thing they don't do is prepare a baby to take his SATs. "Newborns attend to speech stimulation, but they aren't learning words in a way that has long-term consequences," says Fifer. "They could perhaps respond to a pattern heard over and over, like a soap-opera theme their mothers listened to every day, but these 'baby universities' that claim to teach words and music to a fetus are absurd distortions of research. You'd never think of putting a tape recorder blasting Mozart in your sleeping baby's crib, but that's what these people are doing to fetuses."

The question of why the fetus's sense of hearing is so sophisticated long before that seems particularly useful intrigues developmental scientists. "You have to watch the question of adaptiveness, or you can end up telling 'Just So' stories," says Turkewitz. "Because the environment doesn't drive development, the fetus doesn't hear just because there's sound in the womb. Having said that, early on, there are only so many kinds of sound the fetus is exposed to. At first, when it is as flaccid as an underinflated balloon, the womb dampers external sound. Later, when it resembles a taut, blown-up balloon, it acts like an amplifier instead. When its olfactory receptors are stimulated by substances in the amniotic fluid, the fetus changes its rate of swallowing, which also affects volume. These structural alterations mean that in early pregnancy, the fetus mostly hears internal sounds, like heartbeats. As the pregnancy proceeds, it gradually tunes in on voices and other echoes of the world it will soon enter. There isn't much other competing sensory stimulation at the time, so it's possible that early on, hearing might help fetal development by serving as a stimulus around which it can organize. At a certain point, the tactile sense may even assist in this process by allowing the fetus to feel massaged when its mother speaks—one of those scaffolding things that evolution favors."

During the brain's long evolution, the changing acoustics of the womb probably came to influence males and females differently. The brain's right hemisphere starts to mature first, then the left catches up, and at a more rapid rate. This means that early in pregnancy, when noise prevails, the more advanced right hemisphere develops a specialization in handling that kind of sound. Later, when speech penetrates the womb, the right hemisphere is busy handling noise, so the by-now more advanced left hemisphere takes care of voices. As the world of sound changes during pregnancy, females, who mature faster, are at a different stage of development from males. "Therefore, it makes sense, although it hasn't yet been proved, that female brains are less lateralized than male, and better at dealing with language," says Turkewitz. "Whatever role male or female genes play in such differences in brain and behavior, they do so only in the context of the uterine environment, which determines the outcome as much as genes do."

Pressed for a reason for the early development of hearing, scientists may speculate about neurological organization, but most of us would probably suggest a more obvious possibility. Born responsive to his mother's voice, a baby is equipped to help maintain their bond even when they are not in sight of each other—arguably one of his most important tasks. A piece of animal research demonstrates how important the consequences of that aural-vocal bond can be for later behavior as well. When observing two species of identical-looking voles with very different social styles, NIMH researcher Larry Shapiro found that species A voles, who form very strong bonds with their mates and offspring, stay within strict territorial boundaries; species B voles, however, are distant with each other and tend to wander. Shapiro found that although species A pups emit ultrasonic distress sounds when separated from their mothers, species B pups don't. The acoustic environment mediates the social and territorial behavior of both species.

Already attuned to the cycles of day and night and the sounds of his mother's voice and her world, the full-term fetus is poised for the spectacular environmental transition to come. "Our lives don't suddenly begin when we're born," says Turkewitz. "There is a degree of continuity between fetal life and life outside, and changes in our earliest environment cause changes in us, just as our later experiences do."

After his birth in a small-town hospital, Steven was rushed to a larger institution for an initial operation on a heart defect, then flown to Columbia-Presbyterian Medical Center in New York City for additional surgery. He's a pretty baby, rosy and spunky despite the giant incision that bisects his chest. Each day, his mother borrows a car and drives eighty miles to watch over her child. Jane hasn't had a fancy education, but she's an encyclopedia of preemie science, expounding knowledgeably on every aspect of Steven's condition and care. When his monitor continues to go off despite the nurse's assurances that he is okay, she tries not to pester for another checkup but can't help herself. She hovers near the isolette, retreats, circles, penetrates it as best she can, unable to keep her hand out of the porthole for more than a few minutes. The one thing she can't stop worrying about, Jane confesses, is "bonding," a popular unscientific term for the parent and newborn's getting-to-know-you process.

Since the 1960s, enthusiastic new parents have come to consider the birth environment, from the ambience of the delivery room to rituals such as immediate nursing, to be important stimuli to their baby's development. Scientists with an intricate knowledge of what a peculiar hybrid the newborn really is could console Jane by assuring her that because the neonate's nervous system is organized so differently from ours, Jane has missed out on far more than Steven has. A quick survey of the newborn's odd talents suggests something of the magnitude of the environmental transition he has just made, one that is almost as dramatic as arriving on a different planet. He can suck and swallow while breathing through the nose, for example, and when supported even do a kind of reflex "walking," an ability that Myron Hofer suspects allowed him to pace up the uterine wall to get into the preferred head-down birth position. His capacity to perform these neat tricks from the old days in the womb will soon fade in an environment that requires different talents. At birth, his lungs suddenly fill with air and he must breathe; his heat and nutrients are no longer supplied directly by his mother but must be generated from his own tissues. His ability to function in the new cold, bright, airy world is triggered neither by the legendary spank nor his first breath, but by a grueling trip from the sheltered world of the womb.

Birth, the maximum level of stress a human being can withstand, would easily kill an adult. When labor begins, the infant is squeezed with incredible force by the contractions of his formerly cozy home, which also happens to be the most powerful muscle in the human body, including a heavyweight's bicep. Each uterine crunch shuts off the baby's blood supply, lowering his oxygen level. This combination of being squeezed and asphyxiated, which would render us brain-dead, produces an outpouring of the stress hormones adrenaline and noradrenaline at levels twenty times higher than even the laboring mother's stratospheric balance. We are used to thinking of stress as a bad thing, but at birth, it serves a benign purpose, one that may even have lifelong consequences for health. According to Theodore Slotkin, professor of pharmacology, psychiatry, and neurobiology at Duke University Medical Center, it is possible that the immune systems of babies delivered by cesarean section not preceded by labor may fail to activate properly and remain impaired.

"The newborn is just incredibly juiced," says Slotkin. "He's experiencing an extreme version of the flight-or-fight response that occurs when you're as scared as you've ever been. But birth is not an instance in which an anxious physiology causes an anxious psychology. There are no connections between the brain and most of the body's peripheral organs. Even if the baby felt scared, his brain couldn't tell his heart, lungs, or adrenal glands, because it only develops those connections and suppresses the peripheral systems' independent control later on. When the brain matures and takes over those functions, it will need more oxygen for its increased work-load, but its immaturity keeps the newborn from being brain-damaged by oxygen deprivation at birth. It's nature's way of protecting us in the transition from womb to world." According to Hofer, the brain's immaturity also protects us from so-called birth trauma, a concept popularized by psychoanalyst Otto Rank. "Certainly a baby's experience of birth would be very different from ours, because he can't compare things as we can," he says. "Inner experience, or the awareness of a 'me' who is behaving, comes very late in development. Around six months, I suspect maybe a baby has a little. A newborn probably has almost none."

Considering the limited capacity of the newborn's brain and the unique ordeal he has just survived, the decor of the labor room and the immediate performance of various bonding ceremonies are of little importance, to him at any rate. "After being squeezed till his headbones overlap and asphyxiated for sixteen hours, how bright the lights are in the delivery room are the least of his worries," says Slotkin. "Even if such things did matter, the music that usually gets played, for example, is wrong in terms of the sound frequencies babies prefer. These 'gentle' delivery rooms are designed for parents."

Based on his NIMH studies of the process in animals, however, Thomas Insel thinks that "bonding *is* important for the mother. The hormones released at birth enable the imprinting of a special kind of memory that ensures her tie to the infant. It's not an ordinary moment." William Fifer observes that although the data to support the bonding concept "weren't strong to begin with and haven't held up too well, it caused some good things, such as including the father in the delivery. But the whole business got way, way overstated, until parents were convinced that the hour after birth was the most important part of life, and if they didn't get to touch their baby, they all just missed out on it. Somehow, we ended up with an epoxy theory of bonding."

Although a trendy birth-day bonding environment is not essential to sound development, a long-term process that child psychiatrist Daniel Stern calls the "dance of attunement" between mother and baby is. During pedestrian events such as nursing or downing a bottle, an infant also drinks in his mother's touch, smell, voice, warmth, smile, and movements—a sensory feast as vital as food to his development. As an equal partner in the dance, the baby guides the pair's level of emotional intensity; over 90 percent of the time, for example, he is the one to break eye contact. Five minutes in an NICU is long enough for an observer to see that it isn't easy even for a dedicated parent like Jane to partner a preemie. The combination of medical imperatives, other people's expertise, and the immaturity of infants unable to provide the captivating responses we expect deprives both mothers and babies of the comfort of each other's simple touches, sounds, and movements. The waves of frustration eddying through the NICU help explain the fact that although almost all parents initially request that everything be done to preserve the life of the most marginally viable preemie, some give up soon after and stop visiting.

Although it has already improved the lives of babies in some NICUs, Myron Hofer's meticulous animal research on the nature of the mother-infant tie has broader implications. By showing that the integers of "maternal regulation"—a mother's nestling, crooning, rocking, and just plain being there—are the sensory building blocks of our most important bond, he is redefining the nature of mammalian relationships as being not merely social or emotional, but environmental as well.

8

.

THE STRUCTURE
OF EMOTION

I N *As You Like It,* Silvius talks about the experience of love. "It is to be all made of sighs and tears. . . . / It is to be all made of faith and service. . . . / All made of passion, and all made of wishes;/ All adoration, duty, and observance,/ All humbleness, all patience, and impatience;/ All purity, all trial, all obeisance." Shakespeare's words could describe the primal passion of mother and baby as well as the romantic love that it prefigures. As Myron Hofer's work demonstrates, the tie that binds this pair, like that between lovers, is forged by sensory links. Concerning the evolutionary context of emotion, he writes that the "mutual [physiological] regulation between infants and mothers originates before psychological attachment develops, with its mental representations and highly differentiated affective states. Indeed, regulatory processes are the precursors of psychological attachment and its associated emotions. Thus, the neural substrates for emotion are regulated by specific aspects of the mother/infant interaction even before emotional expression becomes clearly differentiated and readily recognizable. It is no small wonder then that many emotions are so strongly affected by social interactions, that certain interactions may be considered to be regulated by relationships, even later in life, and that self-regulation of emotion appears to be intimately tied to the concept of self, a form of internal symbolic relationship."

Although our psychologizing adult sensibility makes us resist thinking about a social relationship as an environment, a mother provides her baby with all the things a physical setting does, including tactile, vestibular, olfactory, auditory, and visual stimuli. For most of human history and in nontechnological cultures still, babies have literally lived on their mothers' bodies, relying on their flesh and blood to keep them warm, fed, dry, safe, and stimulated much as we depend on our homes. The course of Hofer's research on the physiological, behavioral, and evolutionary integers of this archetypal social relationship was set in the early 1970s, when he decided to analyze the so-called "separation response" of an infant mammal isolated from its mother. Writing of this prototype of our sense of loss and grief, the evolutionary psychobiologist Paul MacLean observed: "Separation of the offspring from the mother is calamitous. Perhaps we can trace to this situation the evolutionary roots of unity of the family, unity of the clan, unity of larger societies, as well as the human philosophic yearning for an abstract kind of unity. When mammals opted for a family way of life, they set the stage for one of the most distressful forms of suffering."

It was Hofer's clinical experience with suffering, gained while practicing psychiatry, that first interested him in separation, a subject then off the beaten track of biological research. "To the distressed individual, loss feels like a trauma, but to the doctor treating him, it feels like a mystery that must be addressed," he says. "How does this 'psychological' experience get translated into all those changes not only in mood but also in things like sleep, appetite, concentration? The old theories about separation were of the top-down type, starting with higher processes such as cognition, but a bottom-up process goes on, too. I didn't really see that until I began to study separation in animals."

The traditional top-down view of the human mother-infant bond is expressed by the "social attachment" theory of researchers John Bowlby and Mary Ainsworth. After much observation, the team concluded that this link is emotional in nature, and that being subjected to or spared the distress of separation plays a decisive role in a baby's development. A hard-edged biologist would not deny that the earliest relationship, even among animals, has a powerful social component. If, for example, infants from a naturally aggressive strain of mice are given to easygoing foster mothers to rear, they will grow up to be gentle. However, once psychologist Harry Harlow showed that simple additions to the environment, such as padding wrapped around a wire feeding apparatus, protected young monkeys reared in sensory deprivation from severe behavioral abnormalities, biologically minded scientists began to suspect that a mother contributes more to her infant's survival and development than food, safety, and even love.

As Hofer began to look into this idea, he observed that like other infants, the rat first reacts to isolation from its mother with an acute "protest" of vocalization and agitation. This response is slowly replaced by a much longer-lasting state of "despair," characterized by inactivity, poor appetite, and other expressions of what looks like sadness. Scientists had regarded these short- and long-term reactions to separation as two phases of a single adaptive response to the experience of loss. The vigorous protest stage attracts the mother to return. If she doesn't, the lethargic despair sequel kicks in, helping the infant to avoid the attention of predators and conserve its metabolic resources.

Instead of finding support for this perfectly sensible theory, however, Hofer discovered that protest and despair were not two steps of one response, but independent processes with discrete physiological causes. While an infant's protest reaction could be calmed by exposure to a familiar sensory stimulus, say, the presence of a littermate, its symptoms of despair persisted, not unfolding from protest, but developing on their own. Moreover, a single maternal stimulus could entirely prevent one of the isolated infant's despair symptoms without affecting others. After initial protest, for example, an infant separated from its mother curls up and tunes out. Hofer found that this lethargy was not a global response to missing its dam, but a reaction to the loss of her body heat; when an isolated pup was artificially warmed, it stayed frisky, even though other symptoms of despair persisted. "Science knew that a mother has to keep the baby warm in order for its body and brain to mature," he says, "but no one had any idea that thermal contact with the mother regulates the infant's behavior and activity as well."

Just as the mother regulates the infant's physiology and behavior, Hofer's team found that the offspring get their mother hooked on their relationship and end up running her life, too. Even Lady Macbeth could say, "I have given suck, and know/ How tender 'tis to love the babe that milks me," and the seductive nursing environment offers several physiological motivations for maternal enthusiasm. For example, the pups' suckling causes the mother as well as babies to nod off; her slow-wave sleep in turn signals the release of oxytocin, which lets down her milk and wakes the whole family for another round of nursing. Like a regimen of stimulants and sleeping pills, the pups regulate their mother's activity cycle.

Nor does physiological regulation occur only between mothers and infants. Stephen Suomi, a research psychologist at the NIH who studies the relationships between temperament, environment, and stress in primates, inadvertently discovered a poignant illustration of the benefits of not segregating his subjects according to age. "When a pair of 'grandparents' is quartered with a group of combative young male monkeys," he says, "the old male keeps peace among the boys, and the old female nurtures the more high-strung youngsters. Monkeys only live to about twenty in the wild, and these pairs are nearer thirty—they're thin, gaunt, and balding. But after moving in with the kids, they put on weight and grow their hair back. The males' sperm counts go back up—one even became a father after a ten-year lapse."

During early development, certain forms of nonmaternal regulation can have similarly impressive effects on the young. Female fetuses, for example, are masculinized by prenatal exposure to their male littermates' hormones. And when raised with males, female rats enter puberty considerably earlier than those exposed only to other females; the males' urine contains pheromones, scented chemicals that influence behavior, which appear to enter the young female's brain through her olfactory membrane and affect the release of the pituitary hormones that turn on puberty. "This kind of environmental influence on the timing of the way an organism's genes are expressed can alter its development and account for some of the differences between individuals," says Hofer. "In humans, similar chemical communication through scents during the early parent-child relationship could also exert behavioral effects much later during adolescence."

After hundreds of exacting experiments, Hofer has determined that within the mother-infant relationship, what seems like a single physical function, such as grooming or nursing, is a kind of umbrella that covers tactile, vestibular, olfactory, auditory, and visual stimuli, each of which has a very particular effect on the infant. Pushing beyond the negative influence of separation on development, he has moved into its private realm of sensory stimulation, constructed by mother and infant from numberless exchanges of cues "hidden" within their larger interactions. Through these sensory maternal regulators, he discovered, a mother very precisely controls each element of her infant's physiology, from his cardiovascular rate to his growth hormones, his appetite to his activity level. Her simple presence not only ensures his immediate well-being but also creates a kind of invisible hothouse within which his development can unfold. This biological state in which the pair lives has much in common with addiction. When parted from his mother, an infant doesn't just *miss* her. Not unlike a heroin addict who goes cold turkey, he experiences a physical as well as psychological withdrawal from a host of her sensory stimuli. "What looks like the baby's global reaction of distress over the mother's absence only seems that way because the symptoms of deprivation usually occur together, when the mother's withdrawal activates each one simultaneously," says Hofer. "For a baby, the environment *is* the mother, but a very different mother than the one we're accustomed to thinking of."

The degree to which it is possible to extrapolate from research conducted with animals to human behavior is often debated, and Hofer thinks there are both good and bad reasons for this. "Some people irrationally like to think there's a quantum leap between the human and animal realms, because they find it threatening to think of ourselves as 'just animals,' " he says. "But there are some valid scientific reasons to be careful about drawing conclusions. If even rats and mice sometimes respond differently to the same stimulus, it's certainly not possible to generalize between rodents and people with impunity. There are many parallels, but you have to be careful and to remember that you're never going to answer basic questions about human nature by studying animals. The main products of research are ideas, not answers. Animal research gives you ideas about methods and principles that will allow you to do a few very targeted human experiments. If you get an unequivocal answer to a behavioral question from animal work, and the principle seems to follow in clinical studies as well, then you have a pretty strong case."

Hofer's current work on the genesis of anxiety is a good example of this convincing confluence of animal and clinical research. When a rat pup is separated from its mother, it emits an ultrasonic vocalization whose rate and intensity are related to its temperature. A pup makes more noise when it is cold, even when unconscious and deeply hypothermic, then when it is warm. In a series of ongoing investigations, Hofer is exploring the possibility that the work of crying in response to cold may have evolved to keep the separated infant warm until its mother's return. To test the theory, Hofer severs a pup's vocal nerve and implants a microscopic device to monitor temperature. A similar operation is performed on a control animal from the same litter; the only difference in the two procedures is that the vocal nerve of the control isn't cut. Later, both pups will be removed from the nest and exposed to cold. If Hofer's premise is right, a number of these exercises will show that rat pups that can't vocalize will have more trouble maintaining body temperature.

"That expression people use—'hair standing on end'—describes a thermal regulation response as well as an emotion," he says. "Anxiety might have begun as a response to keep us warm. Its physiology may derive from simpler body responses that got taken over by this thing we call emotion. A few years ago, I wouldn't have said that the infant rat's initial responses to separation, such as vocalization, were the first manifestations of anxiety, as we know it. I wouldn't have talked about rats having anxiety, and I'm not even sure we can now. But a couple of things push me to think that maybe indeed they do." Hofer observes that every drug that is effective in treating human anxiety has a similar effect on the vocalization of distressed infant rats; moreover, drugs that produce disintegrative anxiety in humans increase the vocalization as well, and also distress adult rats. "Again and again," he says, "you find that rats and humans are affected the same way. The clinical data also point to a continuity between an individual's separation problems in childhood and later anxiety, particularly panic attacks. All these different bits of evidence support the idea of anxiety as an entity that develops over time, from the first manifestations of distress at separation in early life to the onset of a full-blown disorder in adulthood."

As these searches for the roots of anxiety highlight, our early milieu has profound implications for our long-term development and well-being. A while ago, a serendipitous accident in Hofer's lab started his team on an investigation of one of the early environmental underpinnings of physical health. A mother rat escaped from her cage and spent the night away from her pups; the next morning, the researchers found the infants' cardiac rates were half their normal level (the resting heart rate of infant animals, including babies, is about twice as high as adults'). Several experiments later, the team found that the mother controls the infant's heart rate with the amount of milk she supplies. The arrival of food causes receptors in the baby's gut to send signals directly to the brain that tell it how to gear heart function. The more milk consumed, the higher the heart rate, and the more nutrient-rich blood distributed to growing body tissues (later in life, when growth slows and this link is no longer needed, the heart rate decreases to adult levels). Over time, deprivation of this hidden regulator would covertly stunt the infant's growth and development. The happy accident that led to this finding has inspired ongoing studies of the links between maternal regulation and cardiovascular development, hypertension, and early obesity. "There's a tremendous variability in individuals' stress responses and their incidence of related diseases—the range between people who give ulcers and those who get them," says Hofer. "This variety must be a function of expressions of different genetic potential played out in different environments."

By showing that the infant is shaped by the ongoing positive bond with its mother, not just by the negative consequences of their separations, Hofer's radical new model of our primal relationship has broken the ideological stranglehold of attachment theory and created a deep intellectual excitement among the younger generation of neuroscientists searching out the origins of behavior. "The issue is no longer *if* a mother is important to her baby, but *how*," says Thomas Insel. "Maternal regulation offers a whole new way of understanding early development and the importance of the environment. We've moved beyond the obsolete approach of social attachment to the realization that the mother not only provides a relationship but also forms an environmental unit with her infant, modifying his physical and behavioral development in very specific ways."

Scientists trying to replicate something of that environmental unit for the benefit of infants separated from their mothers are already putting Hofer's elegant theories to work. So far, the principles of maternal regulation have been employed primarily in NICUs, but in the future, the babies of the majority of American mothers who work outside the home may profit from them as well. At first glance, Hofer's studies seem to suggest that all mothers should follow the lead of !Kung women in Africa and wear their babies in slings nonstop for the first two years. Not so, he says. "Now that we know that separation isn't a single stress response, we'll be able to identify each of its sensory components, which could be provided by other caretakers in the mother's absence. Although maternal regulators ensure the long-range developmental plan in mammals, they don't have to be provided solely by the mother among complex primates. Given time together and closeness, some males will show all kinds of crazy, 'motherly' behavior, probably because they enter a state—in the sense that 'in love' is one—brought about by chemical, maybe even structural, changes. They're really quite wonderful, but perhaps I just say that as a male."

Male or female, anyone who has soothed a howling infant and bystanders' jangled nerves by simply picking up the child and popping him onto a shoulder has successfully exploited the principles of hidden regulation. No matter if the cries were caused by hunger or cold—like magic, the racket stops, and the baby perks up and looks around in an irresistible fashion that reinforces his rescuer's good intentions. Until recently, the assumption, shared by scientists and parents alike, had been that the magic touch in this transaction is provided by touch itself.

Touch, our most basic means of relating to our own bodies and the world around them, is far more essential to survival than vision or hearing. Babies otherwise well cared for in institutions once died of "skin hunger," or inadequate physical contact, and even older children raised in disturbed environments and deprived of loving contact can literally stop growing; these "psychosocial dwarfs" sprout again only in a nurturing setting, such as under the wing of a kindly nurse. The underpinnings of these reactions were partly exposed by a series of regulation experiments done by Saul Schanberg, a researcher at Duke, who found that a mother rat's licking and grooming of her pups literally controls their growth. When a pup was isolated from her ministrations for as little as an hour, its level of growth hormones and appetite fell, only to return to normal at reunion— or when a researcher played mother and stroked the pup firmly with a paintbrush. Working with preemies at the University of Miami, psychologist Tiffany Field made a similar discovery; those massaged for fifteen minutes three times a day gained weight 47 percent faster, were more responsive, and were discharged from the hospital earlier than those who were not.

Well aware of the vital importance of touch, Anneliese Korner nonetheless suspected that there was more than met the eye in the picture of the happy babe in arms. Just as Hofer had done with his rats, she pulled out all the components of this apparently simple, hundred-times-a-day interaction, then tested modified versions on babies. She found that just holding an infant without moving him had little effect on his state of alertness, nor was it nearly as soothing. The best thing about getting picked up was not being touched, but being moved about.

Because adults don't spend much time being jiggled and seem none the worse for it, the regulating effects of vestibular stimulation— rhythm and movement—are hidden from our eyes. But like a sailor on the high seas, the fetus lives in almost constant motion. Even during the night, it responds to the gestures of its restless mother. Its own activity too is necessary to the formation of its nervous system, joints, bones, muscles, lungs, and other body parts. After birth, developing infants continue to need movement to mature their nervous systems and to help them regulate their behavior; among Harlow's monkeys, those that fared best of all had access to a padded apparatus that swung them, and a later study conducted by Korner and Evelyn Thoman showed that rat pups who are rotated weigh more and develop faster than those who are not. Yet the crucial contribution a mother's movements and gestures make to her baby's developmental milieu has remained largely unacknowledged, mostly because their regulatory effects are concealed under the guise of touch.

Some of the babies most deprived of vestibular stimulation benefit from resting on a waterbed mattress designed by Korner to oscillate gently and respond to the baby's motions. Infants in NICUs equipped with these waterbeds are less irritable, suffer fewer episodes of apnea, and spend more time in quiet sleep. "Being on the waterbed for ten days may not change the preemie's life outcome, but it will provide her with a more comfortable ten days," she says. "No one would argue with trying to provide an adult with a drugless good night's rest." Other preemies nestle up to a particularly fetching surrogate maternal regulator: Thoman's preemie-sized blue toy bear, which "breathes" through a special pump in a very regular, quiet-sleep pattern that can be matched to each baby's own respiration. "What we know about preemies suggests that rhythmic stimulation is the most important kind for them," she says. "They get assaulted a lot, but they don't get much movement. This form of stimulation, which is voluntary and customized for each individual, helps them become more neurologically mature. We put the bear in the isolette, and the baby bumps into it by accident at first—unlike other newborns, preemies move around a lot in bed because they lack inhibitory controls. Soon they begin to spend more and more time with the bear, pressing up against it with their heads, torsos, limbs. They treat the bear like a married person treats a spouse—you cuddle up in bed, but maybe not all night."

This reference to one of the homely comforts of marriage hints at the role sensorimotor regulation probably plays in other intimate relationships. One example, reported in *Nature* by researcher Barbara McClintock, is the tendency of the menstrual cycles of young women who live together for a time to synchronize; in animals, this phenomenon is mediated by the olfactory system, and it seems likely that smell has a significant if subtle role in human interactions as well. There is one particular adult relationship about which Hofer is willing to speculate. "In grief reactions, particularly after the death of a spouse, the mourner experiences waves of emotional distress," he says. "Like the protest reaction in young isolated animals, these acute symptoms may be quite different and separate from his or her more chronic disturbances, such as the changes in sleep and diet patterns, that resemble infants' despair responses and contribute to the misery of grief. Evidence suggests that it would be wise to look carefully not just at who, but at what, has been lost. The widowed person is not just suffering from the global psychological distress of missing a spouse, but from the deprivation of the thermal, olfactory, and tactile sensory stimuli of a lifelong interaction. Some of the symptoms of grief could be caused by the sudden withdrawal of multiple regulators woven into the fabric of a long relationship at the biological as well as psychological level."

9

.

ENVIRONMENTAL
ADDICTIONS

L IKE OUR intimate social bonds, particularly the first, our relationship with the larger world is built from countless sensory
interactions between us and our settings. In a very real sense, the
places in our lives get "under our skin" and influence our behavior
in ways that we often don't suspect.

In the 1960s, an era of experimental thinking, Roger Barker had
a particularly wild idea. He decided to chronicle entire days in the
lives of children, recording all their interactions not only with people
but with places and things. After examining his data, the psychologist
came to a startling and most un-American conclusion: their settings
were more important determinants of his subjects' behavior than their
personalities.

Broadening the study of what he called "psychological ecology,"
Barker went on to analyze and contrast the doings of entire small
towns in both Kansas and Yorkshire, England. The more he watched
all sorts of people go about their business in shops, playing fields,
offices, churches, and bars, the more certain he became that individuals and their inanimate surroundings together create systems of

a higher order that take on a life of their own. When we enter one of Barker's "behavior settings"—a school, restaurant, gas station, hospital—everything in that environment encourages us to maintain the status quo. In a sense, we are no longer quirky individuals, but teachers and students, proprietors and customers, doctors and patients. Simply by the way it positions its hours, displays its merchandise, and situates the vendor, even a corner newsstand determines that business will be transacted in a predictable way: just about any clerk or customer will do the right thing. For similar reasons, this year's second grade at P.S. 6 functions pretty much the same as last year's, even though the kids and maybe even the teacher are different.

We unconsciously rely on this Platonic dualism of behavior settings to supply much of the stability of our social institutions. Under its influence, we line up to buy movie tickets rather than clubbing our way to the window, stop at red lights, and lower our voices in libraries, museums, and churches. Sometimes we even gang up with others to enforce a setting's rules, as when we join in shushing whisperers in theaters or giving the cold shoulder to the neighbor who doesn't keep up his yard. Even as small children, we work hard to secure the stability of our settings, and later in life, we're particularly eager to do so in the workplace. In an office that's temporarily understaffed, for example, Barker discovered that most employees will do whatever is required to maintain the established order of things, including working more themselves and upbraiding those who won't. When it comes to business as usual, behavior settings are the *sine qua non*.

On the other hand, behavior settings have a dark side that we brush up against whenever we contemplate changing the positioning of the furniture in the living room, say, or try to get the people we live with to hang their coats in the closet instead of dumping them on a chair. We find that what started out as "a way" has somehow turned into "the way," becoming so entrenched that otherwise competent people are reduced to paying a professional to find a better spot for the piano or to bribing their children to use hangers. It seems that once the environmental particulars of a *modus operandi* work their way into the nervous system, they help close our minds to better options and incline us toward knee-jerk reactions. This insidious tendency to accept familiar behavior settings without question ensures that, for example, no matter how many people cogitate over how a community's adolescents should be educated, the solution is invariably a big regional high school, rather than another possibility, such as several smaller, decentralized buildings, that research shows better serve children.

The more we experience a behavior setting, the greater its power to alter our perception of the "real world." If a child puts on a pair of eyeglasses fitted with special lenses that are flat on one side and wavy on the other, the skewed information delivered to his retina will disorient him, even in a familiar setting. When an adult dons the same glasses, however, he finds that the world looks quite normal, because his brain quickly reorganizes the aberrant input into what he *expects* to see. "Sometimes when a familiar environment has been changed—your office has been painted while you were on vacation—you don't even notice, because you correct for it," says Myron Hofer. "This same tendency can make it hard to see anything new in a scientific experiment. We find confusion hard to tolerate, but it allows us to see things the way they really are. One thing that distinguishes artists may be the retention of that childhood ability to see the world afresh."

On the other hand, our ability to put mind over matter when it comes to our settings gives us some powerful advantages. In one experiment, a person is rotated in the dark for several hours and told to imagine an object rotating with him; responding to this suggestion, his brain will keep his eyes focused on the thing's supposed location. This skill at internalizing environmental cues and settings enables athletes trained in visualization techniques to improve their performances by going through imaginary sports events in their minds, "sensing up" as much as psyching up. That we can not only reorganize "wrong" sensory input into the "right" setting but also change our unconscious physiological reactions by imagining a setting puts a new twist on an old metaphysical question: What *is* the real world?

One reason we work so hard to keep our surroundings predictable is that we rely on them to help us segue smoothly from role to role throughout the day. The Bauhaus furnishings of the office help shore up the persona of the steely executive, who might later turn into a breezy jock at the health club, a Casanova at a bar, and a cozy homebody in his kitchen. Some people, however, are far more dependent on the stabilizing influence of behavior settings than others, and research on their environmental lability sheds light on why places are such powerful influences on the rest of us as well.

Susan, a woman suffering from multiple personality disorder (MPD), worked as a medical technician, but the necessary skills to do this job were not shared by all her alter egos. Not true personalities, these discrete psychological and physiological states of being had first developed to protect her from the memory of abuse in childhood. Usually, the work setting evoked the right "alter," but if it did not, or if an incompetent alter unexpectedly "came out" in the course of the day, Susan had to feign illness or hide until a knowledgeable persona arrived and got the job done.

Before we develop the sturdy, fully formed egos we like to imagine we were born with, our inner lives consist of loose collections of states—constellations of particular physiological, cognitive, and emotional variables that continually occur together in certain contexts. In his mother's arms or his own crib, for example, a baby's heart rate and other metabolic measures slow and he feels content, while a stranger's embrace or an alien bed produces unpleasant arousal and a speeded-up metabolism. In time, the baby learns to associate certain settings and cues with certain states, and the mere sight of mother, highchair, or park initiates physiological and psychological changes. As he matures, his limited palette of states, such as sleep, crying, and alert wakefulness, expands until eventually he experiences different emotions, estimated by developmental scientists to range from four to twenty-seven, that give his personality its unique color. Throughout childhood, he also learns to match his states— sleep, excitement, concentration—to his places—bed, the playground, school—and does the hard work of consolidating his characteristic states into what he and others recognize as his identity.

In children such as Susan, that process of consolidating a consistent ego that prevails regardless of the setting is fractured. To survive the nightmare of repeated abuse, she learned to dissociate, or wall off, various experiences from her consciousness. We all do this from time to time, as when we "space out" to get through a boring commute or lecture, or perhaps a stint in the dentist's chair, but this reaction plays an exaggerated role in the lives of chronically traumatized children. Because dissociation can spare children the awareness of being beaten at home when they're sitting in the classroom, that psychic fragmentation can be said to allow them the semblance of a normal life. But those who eventually develop MPD dissociate so often and under such duress that they eventually lose control over what is meant to be a short-term, occasional defense mechanism. Chopped up into discrete units, their states never get the chance to coalesce into a sturdy ego that can function in all settings.

Because certain states or personalities tend to come out in certain settings, victims of MPD depend heavily on their environments to help them maintain something like normal behavior. "Many of these people structure their homes into little corners that are really microenvironments in which different personalities can express themselves," says Frank Putnam, a research psychiatrist at the NIMH who studies dissociative disorders. "A patient might say, 'This area with all the stuffed animals belongs to Sally, and that one with the clay and paints to Jane.' This type of person tends to have a very hard time in hospitals, where there may not be enough privacy and space to create the contexts that bring out those personalities. If that's the case, behavioral disturbances often erupt from what the patient describes as an incredible sense of internal pressure."

If hearing about that pressure uncomfortably reminds us of how we feel when we miss our Saturday turn as Makarova at ballet class at the Y or as John McEnroe on the tennis court, Putnam isn't surprised. He has learned that discussions of MPD make most people feel uneasy, because the bizarre shifts its victims experience through the day are just extreme examples of transitions we all undergo. "We gloss over the changes in ourselves and our abilities in different environments, and by smoothing out these gaps, at least retrospectively, we maintain the idea that we're the same person at home, in the office, in the car," he says. "But on some level we know there's a lot of discontinuity in our lives, and that makes us uncomfortable."

A good bit of that discontinuity depends on a phenomenon called state-dependent learning. If we learn something in one physiological and behavioral condition, say, drunkenness, and are tested on it later when sober, we will perform less well than we would if intoxicated, if we can perform at all. Similarly, students tested on material in the classrooms in which they learned it do better than when they are tested in a new setting. It seems that the place in which we first master information helps recreate the state necessary to retrieve it, probably by stimulating the right emotions, which are very important influences on memory. Musing on this peculiar power of place, a neuroscientist observed that despite all the cities in which he and his wife have lived, they always consider San Francisco to be "home," because that is where their two children were born. The magic they continue to feel when they visit the Bay Area comes from the way their favorite haunts stir up their senses, unlocking a treasure chest of memories and feelings with the keys of stimuli they might not even be aware of—the smell of crayons, the way sunlight falls on a staircase, the sound of rain pattering on the foliage of a certain kind of tree.

The basic principle that links our places and states is simple: a good or bad environment promotes good or bad memories, which inspire a good or bad mood, which inclines us toward good or bad behavior. We needn't even be consciously aware of a pleasant or unpleasant environmental stimulus for it to shape our states. The mere presence of sunlight increases our willingness to help strangers and tip waiters, and people working in a room slowly permeated by the odor of burnt dust lose their appetites, even though they don't notice the smell. On some level, states and places are internal and external versions of each other.

The mechanics of state-dependent learning usually work so well in gearing our behavior to our setting that who we appear to be, say, at the office may suggest very little of who we are at home. Sometimes we even sense this discrepancy ourselves, says Putnam. "If you're out in the yard on Saturday and suddenly get a call asking for advice on a crisis at work, you may have trouble accessing the skills required, even if the solution involves something you know well." Anyone who has to count fingers to figure out a tip in a taxi knows exactly what he's talking about. "In a cab, no matter who or how competent you are, you're not in control of your environment, and often, your state," he says. "Can the guy speak your language? How is he getting to where you want to go? Is he going to hit that bus? The setting puts you in a special taxi state of mind, and suddenly you can't add or subtract worth a damn. We hardly think about these microstates, or question them very much, because they pass so quickly—as you get out, that cab is history. But for a brief interval, caught between places and states, you perhaps 'weren't yourself,' which probably also explains why we often behave oddly on planes. Once you see that the ability to access and use your skills varies a lot with the context, you realize how discontinuous your sense of self and your abilities really are. That's a discomfiting idea for many people, just as multiple personalities are."

Not unlike MPD patients, we rely on settings to help us be aggressive and energetic at work or on the tennis court, but loving and laid-back at home. Most of us even have a special place or two that brings out special dimensions of ourselves—a basement workbench, a hot sports car, a movie-star bathroom. Profiles of artists invariably include a paragraph about the settings and rituals they depend on to help summon up and sustain their creative states. "To function as a writer, you have to do a certain amount of circling around, almost like a dog, before you can settle down to work," says Putnam, who has a book under his belt. "You have to straighten things on the desk, get the coffee cup just so, sharpen the pencils. You're using environmental cues to help you destabilize whatever else is on your mind, get you out of that state, and stabilize the one associated with writing. Once you're there, think how you resent intrusions, like the phone ringing!" That resentment springs from bitter experience that has taught us that states—especially desirable ones—are finite by nature. "Even if you were in the perfect environment and had no time constraints, you couldn't maintain a creative mood forever," says Putnam. "You can really see the way states run down when you look at kids. You may prolong a small child's agreeable state, you may even coax it back for a while, but holding it is a losing battle."

Places aren't the only environmental props we use to proclaim and enhance our states and identity. The novelist Alberto Moravia has noted that during summers at his Italian seaside home, despite the heat and plenty of privacy, he finds that when he wants to read, and especially to write, he can't be naked. "Clothes are a micro-microenvironment that people use to change state all the time," says Putnam. "That's one of the functions of the shopping spree. Hoping to acquire all kinds of good things associated with it in terms of state, a depressed person buys a wonderful dress or tie."

Macabre examples of a different environmental response to an unhappy state abound in the record of Queen Victoria's decades-long grieving for Prince Albert. Upon her consort's untimely death in middle age, the monarch retired from the world to divide her life between a palace next to his tomb and their Highlands love nest, surrounded by the grisly memento mori—from statues of the departed to locks of the loved one's hair—for which her eponymous age is renowned. It doesn't surprise environmentally minded researchers that Victoria remained almost inconsolable into old age. Like other behavioral states, depression requires a coincidence of environmental and biological cues. When a bereaved person such as the queen surrounds herself with old stimuli and avoids the new ones that could distract her from her gloom, both her serotonin level and spirits plummet. "Getting over a habit, whether it's a lost love or a drug, requires environmental deconditioning," says NIMH pharmacology researcher Susan Weiss. "You can't recover if you sit around in your old haunts. To condition yourself to health, you have to break past associations and respond to the positive stimuli of new things and places, which will generate even more good stuff."

Although we usually think of drugs when we hear the word addiction, the term applies to any habitual or compulsive devotion. Much of the strength of that devotion, whether to a person, an intoxicant, or a pursuit, comes from the environments of our past and present, which hover like ghosts beneath the surface of our awareness, haunting us and our behavior. When Vladimir Nabokov came upon E. B. White's definition of "miracle" as "blue snow on a red barn," he was instantly flooded by the Russia he had lost.

In *The Teachings of Don Juan,* one of the first anecdotes concerns the Yacqui shaman telling the author Carlos Castaneda to find his "place" on the porch, a process that required the anthropologist to move all over the floor until he found the spot where he felt most comfortable. The underpinnings of such finely honed perceptions defy analysis, but a new scientific perspective on drug abuse suggests something of the way settings get under our skin, physiologically and psychologically, and help shape our behavior. During the Vietnam War, large numbers of American soldiers overseas started using heroin. In September 1972 alone, 2,000 of these vets came home. Accustomed to the intransigence of the heroin addiction simultaneously ravaging American ghettoes, where an 80 to 90 percent relapse rate prevails, drug experts anticipated a huge social nightmare. It never materialized. Like most patients prescribed large doses of opiates in the hospital, most veterans left heroin behind when they returned to an environment they did not associate with the chemical.

If the drug is the spider of addiction, it more and more seems that the settings in which it is used comprise the web. "We have a lot of intricate data on drugs and what they do to this or that receptor in the brain, but very little of it helps explain their effects on abusers," says Weiss. "There's more to addiction than chemicals and personal psychology, because we can detox someone who is highly motivated to change, and he can still fail. The missing dimension is the environmental factors that precipitate drug use."

The recorded quirks of drug addicts have long hinted that chemicals are not the only factor in getting hooked. "For more than a century, we've realized that if a person knows he is going to take a drug, he has the capacity to resist its effects," says Shepard Siegel, a professor of psychology at McMaster University in Hamilton, Ontario. "It was observed way back in 1859 that 'Man is very much the creature of habit. By drinking regularly at certain times he feels a longing for liquor at the return of those periods . . . and even in certain company or a particular tavern in which he is in the habit of taking his libations.' " Like other addiction researchers, Siegel had noticed that a person's drug experiences are not always determined by his drug: an addict who is unable to buy heroin sometimes injects water and actually feels better, for example, and occasionally an experienced addict takes his usual fix and dies, as if he had overdosed. "After interviewing ten users who had survived this kind of reaction, I found that seven had injected in a strange setting," he says. "The element of novelty varied. One woman, who always took her first fix in the bathroom after her mother went to work, almost died one day when her mother, who actually hadn't left yet, knocked on the door as she was injecting. For others, the difference was just taking the drug in a new neighborhood."

Back in his lab, Siegel gained an important insight into these anomalous drug reactions. He found that when an animal is placed in the environment in which it usually gets a drug, it reacts with the opposite physiological responses to those the drug produces, as if its body were trying to preserve a normal physiological state by combating the drug's imminent influence. In anticipation of the analgesia to come, for example, a rat who gets morphine in a certain place becomes extraordinarily sensitive to pain while there. More research revealed that tolerance—a user's ability to take increasing amounts of a drug without experiencing its effects—occurs because the environments in which the drug is usually taken elicit the very responses that attenuate its effects. When an addict is deprived of the damper of familiar surroundings, his customary dose of heroin can suddenly pack a deadly wallop, just as the glass of wine only rarely taken at lunch in a restaurant goes to one's head far more than the one habitually drunk with dinner at home.

An almost disturbingly simple mechanism explains much of the pull our surroundings exert on our behavior. "If I slice a lemon in front of you, you will salivate," says Siegel. "When two stimuli are paired, we automatically learn the association between them, whether we realize it, or want to make it, or not. That's why your cat jumps on the counter when it hears the can opener, and why Pavlov's dogs salivated at the sound of a bell that usually preceded their meals. Environmental cues teach us what to expect in different places." Some of the combat veterans treated by Bessel van der Kolk, a professor of psychiatry at Harvard, learned to expect such high arousal in the exotic milieu of Vietnam that when they look at magazine photos of jungle warfare or fifteen minutes of a movie such as *Platoon*, their nervous systems pump out a surge of opiates. Thousands of miles and twenty years away from the war, the old terrors and excitement summoned up by mere images anesthetize these victims of so-called posttraumatic stress syndrome like the equivalent of eight milligrams of morphine. Such bursts of natural painkillers, meant to temper occasional stress, may even draw chronically traumatized people, from abused children to combat veterans, to the very stimuli that torment them.

Describing how addiction spreads its tentacles in the brain, Weiss could be talking about the way places covertly lock more pedestrian stimuli into our nervous systems as well: "Our brains are so adapted to make associations with the environment that whether we want to or not, we link our experiences and their settings, and those two things together produce the behavior." It's possible that the links that make us drool at a sliced lemon or even stop for a drink when we pass a favorite watering hole are stored in a special neurological memory system. Our more conscious associations—what we had for breakfast or the name of that nice man at the party—are filed in so-called representational memory, centered in the amygdala and hippocampus. A deeper "habit" memory, based in the striatum and the caudate nucleus, handles routine things that happen all the time, such as making the right twists and turns during our daily commute and stopping at red lights. According to one paradigm, the ground-work of addiction—the getting-to-like-it stage—is laid in the representational memory system. Later, the associations made with drugs move over to the habit system, far less accessible to our conscious interventions. There, all sorts of subliminal cues can trigger the urge to get high, which makes kicking a drug habit difficult even for the well motivated. "Heroin addicts will experience a craving when they smell a burnt match or pass a corner where they are accustomed to buy drugs, even though they're unaware of the reason," says Siegel. "This kind of physiological response to hidden cues accounts for withdrawal symptoms, which don't spring from the toxic effects of the previous drug dose but from anticipation of the next one."

Whatever the fine points of its etiology, the link forged in the brain between settings and drugs is so strong that after a while, the user's whole world seems to whisper "cocaine"—or "gin," or whatever. He walks into a certain hangout frequented by his drug-using friends, and their faces and the music, colors, light, and smells all spell "high." Because his positive associations with these cues are locked into his brain by a potent chemical reinforcer such as cocaine, which animals will self-administer till they drop dead, says Weiss, "eventually the mere sight of his paycheck will prompt a powerful involuntary urge to use, because paychecks mean a binge."

One reason traditional addiction programs have had such a poor track record is that they have largely ignored the role of the environment. "Treatment should involve systematic exposure to drug-related environmental cues without the reinforcement of getting the drug, so the addict's body learns not to give the anticipatory reactions," says Siegel. "The best treatment of all remains the so-called geographical cure. Studies from all over the world show that after a year, most of those who don't relapse after drug treatment have relocated. The new environment may not be drug-free or even seem very different—it could be another ghetto, for example—but it's free of the cues associated with use for that person."

Addiction is just a dramatic example of the exquisite responsiveness to the environment that enables us to function efficiently and survive. "If you enter a setting in which you often eat, your digestive juices are already flowing when you pull up your chair," says Siegel. "If you're an ex-athlete, you experience physiological changes just watching your old sport. Because we're homeostatic machines that can only function if our internal milieu stays within a narrow range of temperature, heart rate, blood sugar, and blood pressure, many of our resources are directed to maintaining this optimum internal environment. And one way we do that is by anticipating and correcting for influences in the external environment that might disturb it."

SYNCHRONY

■

PART III

■

10

.

STIMULATION:
LESS IS MORE,
MORE OR LESS

A NCIENT RUINS, from the "Nazca lines" in Peru to the monoliths of Stonehenge, offer spectacular testimony to the importance long accorded to what Pliny the Elder first called geomancy, or people's propensity for using external signs to figure out what to do next. Although this tradition tended to decline with the rise of modern science and Judeo-Christian religion, some of the practical Chinese continue to rely on the system their ancestors developed to help people find and create settings both portentous and propitious. During the darkest days of the collapse of the boom economy of the 1980s, for example, one Chinese-American financier had several big deals in a row crash around his ears. His stress level soared, and he knew he needed some help. Instead of heading for the treadmill or the medicine chest, he consulted a *feng shui* practitioner, who addressed his client's business reverses by moving around the furnishings in his office.

An eclectic discipline, *feng shui* combines bits of art, geophysical observation, psychology, religion, folklore, and plain common sense. Although it is not science, it incorporates many insights from researchers on the link between our internal states and external environments. Just as *yin* is balanced by *yang,* the female by the male, and the earth by the sky, a place that has good *feng shui* is neither boring nor agitating but promotes the right level of arousal for the business at hand. While this kind of setting attracts us and makes us feel easy in body and mind, spots we tend to avoid or feel uncomfortable in are likely to have problems with their flow of *chi,* or the energy the Chinese believe animates the earth and all living things. Although most Westerners are skeptical of such concepts, an increasing number of architects and designers attempting to balance urbanized settings are finding inspiration in this ancient hybrid method of harmonizing people and places.

"Perhaps the first farmer to plant his crops according to southern exposure was the first person to practice *feng shui,*" says Sarah Rossbach, the author of two books on the subject and a practitioner herself; while in Hong Kong on a fellowship, she learned *feng shui* from Lin Yun, a master who happened to be her language instructor. "The early masters tried to read the earth's pulse in order to find the best places for particular functions, especially the very important siting of ancestors' graves. Indoors, instead of making their calculations according to features in the landscape, they used corners, walls, and windows. *Feng shui* may not always be logical, but the underlying concept, which is that living in harmony with your environment can improve your life, makes sense. Because we do react to, sometimes even mirror, our surroundings, changing them should help us to change ourselves."

Under ideal circumstances, good *feng shui* starts before a building is erected. The owner or architect consults a "doctor," who determines the details of the structure's design and placement that will invite the steady stream of *chi* that helps ensure good fortune. When the Bank of Hong Kong failed soon after it was built, the Chinese were not surprised; the benighted architects hadn't consulted a geomancer, and the structure's *chi* was terrible. In Hollywood, the management of the Creative Artists Agency, which represents the cream of the town's talent, took no such chances. Their new headquarters was *feng shui*'d as soon as it was a gleam in their eye, and in terms of business, so far, so good. In most cases, however, the *feng shui* doctor does not get to apply his art to a *tabula rasa;* he must try to "cure" or "cleanse" buildings and settings that weren't designed with *chi* in mind but, like Topsy, just grew.

Much in the way that an acupuncturist seeks to eliminate obstacles to the flow of energy in his patient's body, the *feng shui* practitioner treats a troubled environment by removing impediments to its *chi*, usually by addressing certain physical features, from a rock in a landscape to the door of a house. Should he decide, for example, that his clients are adversely influenced by their parlor's view of a graveyard or a used-car lot, he might suggest that they hide it behind an opaque screen that still lets in the light. If they tend to avoid a certain dark room or corner, they could make it more appealing with a lamp or plant. Similarly, the right-handed person can move his phone to the right side of his desk, so he doesn't have to twist to answer it, making life seem more difficult than it need be. The general *feng shui* idea is that if a setting doesn't make you feel welcome, tinker with it until it does.

One can learn more about *feng shui* during a few minutes spent watching Rossbach cure a client's generic modern apartment than from plowing through scholarly explications. "Entrances are very important," she says, politely surveying the nondescript vestibule. "The *feng shui* of a place is the quality that strikes you as soon as you enter it. If the first thing you see from the doorway is the kitchen, you'll eat too much. If it's the bathroom, you'll be going all the time. If it's a bedroom, you'll lie down too much. Your entry should emphasize what's important to you. This family has young students at home, so perhaps they might put a small table here, with some books on it. A mirror would be good, too, because visitors could see themselves, smile, fix their hair. A landscape can be nice, too, especially in the city." She frowns at the way the front door lines up with a bedroom door and window across the hall—too much unbroken *chi*: "There should be a crystal or a mirror in that window to disperse the flow. Otherwise, money will fly out."

As she moves through the apartment, many of Rossbach's suggestions concern modifications of the level of stimulation in certain microenvironments. The bedroom is located in what *feng shui* principles have determined to be the layout's "wealth position," which brings good fortune, and the bed happens to be in the "commanding position"—kitty-corner from the door, with the headboard against the back corner. Because the couple can see the entrance to the room while reclining, they won't be startled, a form of overstimulation considered very troublesome by the Chinese. "If the room couldn't accommodate the bed in the commanding position," she says, "the clients could avoid being surprised by hanging a mirror where it would reflect the doorway. For the same reason, they should put a mirror above the stove—if you're startled while cooking, you could get burned, and that underlying anxiety makes you uneasier than need be. Of the cures for architectural problems you can't change, mirrors are the aspirin. They not only ward off bad things but also bring good views, light, space, and extra vantage points." Despite its generally good *feng shui*, the bedroom has a few problems, says Rossbach. A Raggedy Ann doll from the wife's former boyfriend, still lying on her bed, is an immature influence that could keep her hung up on the past. Furthermore, the bed shouldn't sit between a door and a window, which invites drafts and illness, "and a door that sticks a little means an extra effort first thing every morning, which suggests that life is a struggle."

As she leaves her clients' apartment, Rossbach accepts their preordained offering of nine red envelopes holding money. When they ask if she finds it difficult to reconcile her rigorous scholarly background with a practice that incorporates superstition as well as good sense, she smiles. "For thousands of years, the mystical has been a part of *feng shui*. Why try to rationalize it or edit it out? At the very least, it adds anthropological interest. I don't know why *feng shui* works, but I know that it does."

Like a *feng shui* practitioner, psychologist Peter Suedfeld is interested in harmonizing people and their settings. When describing environments, however, he talks of overstimulation and understimulation rather than too much or too little *chi*. After twenty-five years spent studying the reactions of prisoners, submariners, the shipwrecked, and others who have dealt with situations so over- and understimulating that most of us experience them only vicariously in darkened theaters, as well as conducting experiments at the University of British Columbia, he is convinced that when it comes to stimulation levels in the modern world, within the bounds of reason, less is more.

Volunteers for experiments in what Suedfeld calls the "restricted environmental stimulation technique" spend twenty-four hours in solitary, sealed away from the world in a soundproofed chamber that is "darker than any room you've ever been in," as he puts it. The old-fashioned term for this experience is "sensory deprivation," two words guaranteed to call up harrowing prison movies and ominous recollections from Psych 1 concerning the effects of impoverished environments on infant monkeys. "You hear these terrible things about reduced-stimulation environments—that they make you hallucinate or go insane," he says. "But if you look at the literature, the 'hallucinations' are daydreams and fantasies, and the insanity never happens at all. If someone panics and comes screaming out of a 'float tank' or dark, quiet chamber, it's because he wasn't properly prepared for the experience. When people have time to see the environment first and to ask questions, rather than coming out screaming, most want to stay in longer. Those who have particular trouble with highly arousing settings, such as babies born addicted and autistic children, do especially well. In all my years of research, I've never had a screamer, and fewer than five percent of the subjects who've entered a chamber have wanted to stop before twenty-four hours. Most are surprised to learn their time is up."

People can be reluctant to leave reduced-stimulation environments for several reasons. If a dead-quiet, pitch-black room seems to stretch that term, says Suedfeld, it is because we forget that much of the input we get comes from within. When external cues are minimal, memory and creativity take up the slack and often improve. It is the luxury of being able to tune in to this quieter private frequency, usually swamped by external static, that people find so restorative that they will pay $20 an hour to unwind at commercial "float centers." Thousands of years before such things existed, those seeking the same effect turned to meditative disciplines such as yoga. Because sleep is the only escape from overstimulation that many people have, yogis consider it more of a mental than physical requirement; as practitioners become more adept, they learn other ways to quiet the nervous system, and eventually require less time in bed.

Perhaps the most significant benefit people report to Suedfeld following a spell in his low-stimulation environment is the way it fosters behavioral change. "If you add twenty-four hours in a chamber to some training in behavioral management techniques, the relapse rate for those trying to overcome an addiction, say, smoking, drastically drops," he says. "Instead of allowing you to get distracted and tune them out, the chamber induces you to concentrate very hard on the therapeutic stimuli. It also ensures that you have a whole day with nothing to do but introspect, to work something out for yourself. This concept isn't new. Many groups, including the Early Christian hermits and anchorites, have built isolated, retreat-type experiences into their cultures for the same reason."

An hour in a float tank or a day in a dark chamber can be restorative. But people who for one reason or another find themselves in places that even the Desert Fathers would have found hard going can discover that an overly long stay in an understimulating environment poses problems. Extreme reactions such as the "cabin fever" reported during long, dark Alaskan winters and the desert-related hysteria culminating in murder or suicide, called *cafard* by the French Foreign Legion, are probably behavioral breakdowns caused by a long, dull, environmentally enforced confinement. When a boring setting is also dangerous, the combination can be deadly. A lone sailor attempting a round-the-world race committed suicide a while ago because he had cheated, but most *cognoscenti* who heard about the death initially blamed it on his grueling months-long vigil on vast wastes of open sea.

Stuck for too long in an understimulating milieu, we subconsciously try to stir up some action from within, and even exaggerate it. This tendency accounts for the high rates of hypochondria and psychosomatic problems, from insomnia to vague aches and pains, reported in places such as polar research stations and submarines. If a situation is particularly dull and confining—say, a few weeks in a hospital rigged up in traction—our propensity to look inward can generate a highly suggestible state of mind. The line between reality and elaborated, dreamlike thoughts can blur until we start to see, hear, or otherwise sense funny things.

Throughout history, people isolated in the wilderness have reported a feeling of communion with some force greater than the self; for example, Moses, Jesus, and Mohammed all encountered supernatural beings during vigils in the desert. Not just monks and ascetics, however, but hikers, sailors, and young males going through initiation rites have routinely described transcendent sensations of elation, oneness with the universe, and an appreciation for natural grandeur contrasted with their own smallness. The records show that a tenure in the wilds that incorporates other stresses, such as cold or starvation, can evoke a particular type of hallucination. Perfectly sane explorers, climbers, and survivors of plane crashes and shipwrecks have given vivid accounts of sensing, even seeing and conversing with, a benign presence that offers help. Lying ill from food poisoning while making the first solo navigation round the globe, Joshua Slocum was visited by the pilot of the *Pinta,* one of Columbus' ships, who steered his boat for a day and returned to warn of danger later in the voyage. Near Denali's summit, a climber suffering from altitude sickness turned back despite his companions' urgings when he saw a giant image of his wife's face fill the sky, warning him not to go on. Sometimes adventurers isolated in small groups share the same eerie visitation: Antarctic explorer Sir Ernest Shackleton and two companions all had "a curious feeling on the march that there was another person with us."

Climber-scientist Peter Hackett describes a classic sensed-presence experience. "One night I was alone near the summit of Everest," he says. "The rest of the team was camped about five thousand feet below. Suddenly, I saw one of the members come into my tent with an extra oxygen bottle. I was very appreciative when this guy just showed up at three in the morning, because oxygen deficiency is much more extreme when you're trying to sleep and your respiration rate declines. 'God! You're such a great guy!' I said to him. 'Thanks for doing this.' This hallucination was very much like those reports by people who have had a near-death experience in which a guide came out to meet them. That kind of image is a powerful thing, which is probably why visualization exercises often involve seeing someone who's going to help you."

Although a sensed-presence experience would be bizarre in normal settings, Suedfeld finds that it is a normal response to bizarre settings. "When incoming stimuli are severely restricted, we refocus the attention that's usually directed outside and concentrate on our own internal stimuli instead," he says. "If we're in danger, we're also highly motivated to search for any external cues that could help us survive, so we'd be inclined to elaborate on whatever stimuli come from the environment—sounds, shadows, anything. Either or both of these tendencies make a sensed-presence hallucination likelier. We should study this phenomenon further and remove it from the realms of mysticism and psychopathology, because more and more people are spending time in extreme environments. They're going to have these experiences and feel uneasy about them, or at least about reporting them."

Suedfeld's research on and personal experience with what he calls "challenging" environments make him skeptical about the wisdom of cramming ever more people, places, and things into our days, to the point that fatigue has become a leading modern malaise. "Most of us live toward the high end of the stimulation spectrum," he says. "We're programmed by evolution and experience to handle a wide range, but it does have its limits. To say that some of us become addicted to the urban adrenaline rush may be a metaphor, but it's a good one. It's important psychologically and probably neurologically to reduce that load periodically and restore the balance. That's why people who live in big cities generally vacation someplace where they can relax and unwind in an atmosphere that offers relatively low stimulation. It's a cliché, but moderation is important. You don't want to live your life either on the high or low end, but the low isn't nearly as bad as people think. The crews at polar stations spend up to six months in the dark with very little in the way of modern technological stimulation, yet they like the awesome beauty and grandeur of the exotic natural setting, and they also generally like their companions. The lack of human scale—giant icebergs and so on—makes a few uncomfortable, but most love the experience and some even opt to spend their careers there. For that matter, even prisoners in solitary generally do fine."

If a float tank sits at one end of the spectrum of stimulation, New York is at the other. Since the days of Sodom and Gomorrah, cities have been singled out as bad places that make for bad behavior, and people have griped about their noise and dirt, crowds and crime. Some of the reasons some of us love to hate the city are built in: its architecture, for example, restricts sunlight, forms wind tunnels, and traps heat. These troubles don't usually head most city-bashing lists, however. The vexation they cause merely adds to the stress generated by the source of most urban environmental problems: people. In a city, their huge number ensures that it will not only be more crowded than other places, but also more restrictive, competitive, bureaucratic, hectic, and just plain arousing; the quietest times in New Yorkers' apartments are louder than the noisiest in small towns.

Testimony to America's traditional suspicion of the town and bias toward the wide-open spaces, figuratively at least, is as close as the strains from the local country-and-western radio station, and even the most upbeat urbanite must admit that city life occasionally gets him down. There are many theories about what happens to his mind and body when it does. According to an early hypothesis, the city simply provides too much stimulation of almost every kind, bombarding us so relentlessly that in self-protection, we tune out and turn off. The price for this coping strategy can be a reduction in the quantity and quality of our experiences, as well as the erection of social barriers and hierarchies that relegate traditional communal responsibilities to bureaucracies.

However, theories about the origins of urban stress are becoming more complicated. According to one point of view, the city is no more than the sum of its parts; rather than being dealt a single mighty blow of global overstimulation, we're nibbled at by a gaggle of separate stressors. Each form of aggravation, from the racket of a noisy intersection to the strangled sensations of the bus at rush hour, produces its own effects, from impaired concentration to irritability. Our well-being therefore depends on how successfully we deal with individual problems. If we soundproof the apartment, the noise outside no longer distracts us, and if we walk to work rather than ride the bus, we are no longer lost in the shuffle. Still other theories about the roots of urban malaise suggest that the constraints the city imposes on our behavior, such as traffic and crime, are to blame, or the fact that a metropolis is like a vast corporation in which the applicants for jobs and benefits exceed the available resources. Although the experts don't agree on just how the city gets on our nerves, they concur that many of the components of that reaction are incorporated into a ride on the New York subway at eight o'clock on a weekday morning.

The global feeling many commuters experience at that hour on the Seventh Avenue line—the combination of dysphoria and jitters that is the hallmark of unpleasant arousal—is actually the cumulative effect of many noxious influences. When the brain perceives a stimulus, whether it is birdsong in the country or the shriek of car wheels in the city, its reticular activating system, the neural switchboard for processing external and internal feedback, puts the nervous system on alert. Then it is up to us to identify the stimulus that has stirred us up and decide how we feel about it. If we hear a grizzly growl while we are on a hunting trip, for example, we experience the arousal it provokes as excitement—a gain—but if a plane crash has left us stranded in the wilderness, we perceive the excitation sparked by the same sound as fear—a loss. In less dramatic circumstances, even pinning down the stimulus that is rousing our nervous system can be difficult. Waiting for a train at a busy station, lost in thoughts of chores that must be handled once we get to work, we might not consciously notice the static buzzing on the loudspeaker or the reek of urine. Nevertheless, we might become uneasy, and perhaps mistakenly assign this sensation to a gripe with the boss or a problem with an assignment. Even if we successfully zero in on the noise or the foul odor as the source of our upset, we still have some discriminating to do. Should we decide the offensive stimulus is really not significant, we just might shrug and tune out by reading the paper. But if it strikes us as invasive, say, or harmful, or emblematic of urban blight, or if the nastiness is aggravated by a fight in the queue or a delay in service, we are apt to feel the first jangling signs of the "flight-or-fight" response.

As a short-term reaction to an emergency, say, a mugging, this neurochemical jolt, which evolved to help us deal with a sudden threat by running away or doing battle, is extraordinarily useful. But experienced chronically in situations where neither fight nor flight is appropriate or even possible, such as the subway or freeway at rush hour, the changes in blood pressure, respiration rate, hormone levels, muscle tension, and digestion it sets off can take a disastrous toll on well-being. Although we most often associate this stress response with cardiovascular and gastrointestinal problems, lowered immunity, fatigue, and head- and muscle aches, it affects our behavior as well, skewing performance, mood, and sociability. In its shadow, work becomes a chore, our voices move a notch higher, and other people keep their distance.

Subway riders might argue over which aspects of their experience have the most potential for provoking the fight-or-flight response. For starters, there is the aggravation endemic to commuting. On public transit systems, we are not only bereft of an automobile's distractions and amenities, but we are also more vulnerable to the ordeal of long, unpredictable waits that set off waves of helpless anger. And then, there's crowding. While misery may love company elsewhere, sardinelike proximity to strangers on trains and buses kindles an assortment of antisocial feelings from aggressiveness to indifference; this experience is so upsetting that when asked to assess the quality of their commutes, train riders in a Swedish study considered elbow room twice as important as the length of their ride. Some of the techniques we use to reduce the overstimulation caused by crowding can actually make things worse. When passengers try to avoid eye and body contact with each other, for example, an accidental jostling or trivial remark can seem like an act of aggression, sometimes sparking murderous rages unimaginable in less stressful settings.

The *pièce de résistance* of subway strain is noise. For that matter, of all environmental stresses on behavior, it is the worst—an almost entirely man-made plague that rarely occurs in nature. Often to a greater extent than better-publicized environmental problems such as air and water pollution, noise generated by people and their machines causes readily measurable physiological and psychological changes. We find noise so disturbing because it both distracts and restricts. We can read, talk, or think if the subway is hot or crowded, but not if we are sitting next to a blaring radio or wincing at the squeal of brakes. Along with impairing concentration, memory, mood, and performance, noise undermines sociability. New York City assigns extra police to theaters holding concerts of rap music, and an experiment on the effect of noise on aggression supports that decision; subjects primed by a violent film and loud noise punished their "victims" with more "shocks" than a control group, suggesting that incendiary social cues, such as hate-mongering lyrics, coupled with the decibel levels of a rock performance could indeed boost aggression.

Everyone agrees that noise pollution is bad. The chief hurdle to reducing it is defining it. So far, the best efforts in that direction amount to saying that noise is any sound the individual listener doesn't like. Volume can't be the sole standard. Because sound over 90 decibels is physiologically damaging and can cause deafness, the federal occupational safety standard limits workers to no more than thirty minutes' exposure to 110 decibels per day. People who would protest that kind of racket on the job, however, might object vigorously if someone tried to turn down the amps at their favorite nightclub, where music blares at 120 decibels for hours on end. Nor can aesthetics be the final arbiter. We might enjoy Mozart at a volume at which rap would be intolerable, or vice versa; no less sensitive a creature than Vladimir Nabokov found all music unbearable. And most important of all, because the difference between sound and noise depends not only on volume and personal taste but also on predictability and control, the person blowing his horn or mowing his lawn isn't nearly as upset by the din as others in his proximity.

As any patient who has listened to the whine of the dentist's drill knows, exposure to a noxious stimulus is doubly stressful when no escape is possible. Trapped helplessly in a far more hair-raising environment, some soldiers in the POW camps of the Korean war were subject to a strange condition known as "give-it-up-itis": they simply refused to rise in the morning, wash, or work. The source of their apathy seemed to be emotional, because if roused by a slap or a request to help a comrade, most soldiers were able to snap out of it. The Korean War POWs have been unfairly singled out as somehow wimpy, says Peter Suedfeld, not only because many others in similar situations have suffered from give-it-up-itis but also because "in places like those prison camps, it's hard to separate the psychological effects from those of extreme malnutrition or cold, to which some are much more susceptible than others." The benumbed reaction of people stuck in highly arousing situations partly derives from endorphin levels raised by pain and stress, which anesthetize the mind as well as the body. The impact of these natural painkillers evoked by freezing and immobility on feelings of futility probably helps explain the fate of apparently healthy climbers found sitting frozen on Denali's ledges from time to time. "Many of them have been Oriental," says William Mills, who treats many climbers. "Faced with an emergency up there, it's almost as if their cultural reaction were fatalism." Animal experiments on a phenomenon known as "learned helplessness" also provide some insight into this response to an environment: rats who repeatedly fail to escape a painful stimulus will give up and stop trying, even after the option to flee is restored. This sense of being overwhelmed by the outside world is a major ingredient in the recipe for clinical depression.

On the subway the stimulation we are subjected to often meets just about anyone's criteria for stress, being simultaneously unpleasant, intense, unpredictable, and uncontrollable. Under a barrage of maddening rush-hour input, like well-drilled soldiers we react with conditioned responses designed to shield us. But even if we seem to succeed at tuning out over a magazine or mantra—which perhaps helps explain why so many people read prayer books and Scripture underground—insidious environmental stressors such as noise do their damage, sandbagging us after exposure ceases. "We may not be attending to a stimulus, but it's still contributing to our overall arousal level, which is higher on the New York subway than in other places," says Suedfeld. "And in that kind of stressful situation, where neither flight nor fight is appropriate, you just have to sit there and let your muscle tension and hormones build up with no release. Those reactions continue to unfold and create problems later." Although at times they might seem like vast laboratories for the study of stress, he hastens to point out that cities have many wonderful features as well—"the reasons urban communities originally began. Some of those positive characteristics are precisely the stimuli that, when excessive or prolonged, turn negative."

Few places are as overly stimulating as the rush-hour subway, yet some of the less overtly jittery spots where we spend far more time exact their tolls as well. As we move further into the postmodern age of information, the workplace is changing fast, causing occupational safety specialists to focus on problems that would have seemed lightweight to their predecessors. Not long ago, for example, "industrial fatigue" meant the hard-hat exhaustion of steelworkers and coal miners. Now it is just as likely to refer to the weariness, eyestrain, and aches and pains of computer operators who spend long periods with poorly designed VDTs, desks, and chairs. Environmental psychologists have shown that the proper adjustment of a single element at a computer station calls for painstaking microanalysis. To evaluate the screen, for example, one must consider its height, tilt, and distance from the operator, the size and clarity of its characters, and its brightness, glare, and flickering. Because its effects are often subtle, combine with other irritants, and bother us after exposure ceases, trying to pinpoint an environmental stressor is difficult for the layman. As a result, we often end up blaming its noxious influence on something else—the project, the boss, or "stress" in general—thus perpetuating the dilemma.

At work, as in many places, noise leads the list of environmental stressors. Peace and quiet are the major ambient influences on efficiency, but are increasingly rare commodities in offices. When a certain company moved to one of the new open offices that are cheaper to build and operate, everyone was initially excited about the fancy new equipment and pretty walls and carpets of the loftlike space. Without consulting any of the workers about their needs, however, the designers had put eight people, with jobs requiring them both to concentrate for long periods and to be on the phone a lot, in a windowless cluster separated only by flimsy half partitions. When a phone rang, every head jerked up to check whose it was. Because people cannot help paying attention to it, the conversation of others was a far worse problem; a whopping half of all employees say they are disturbed by colleagues' chatter and phone calls. Someone working on a difficult assignment might be forced to eavesdrop on an interview being conducted on his right, a dinner date being made on his left, and several other interchanges in adjacent clusters. Even when quiet prevailed, the workers suffered the consequences of aural pollution; research subjects make more errors and get more frustrated when doing tasks *following* noise exposure.

At least the management of this company didn't pipe in Muzak. Because quieter environments mean higher productivity, some firms decided to broadcast music in their buildings as long ago as the 1940s, hoping pleasant sounds would buffer nasty ones. But different types of sounds, including nice ones, have different effects on different sorts of tasks, and trying to fix such a complex environmental problem with one big Band-Aid is often a wasted effort. The kind of background humming that emanates from a copier, for example, has little effect on clerical or mental chores. Yet it impedes those doing two things at once, or performing tasks that require motor skills and vigilance, probably because it masks other crucial sensory stimuli. On the other hand, irregularly occurring noise, such as that of ringing phones, doesn't interfere with clerical tasks but hampers more difficult mental chores, motor skills, and vigilance. Even the sweet strains of music only really enhance simple jobs requiring attention; by boosting our arousal level enough to overcome monotony, Vivaldi or Willie Nelson might help us do a better job of waxing the floor, but could interfere with harder projects.

In short, whoever is stamping out all these modern cookie-cutter offices foolishly ignores evidence that getting just the right degree of stimulation—enough to skirt boredom on one hand and anxiety on the other—is crucial to optimum performance. Cost-conscious managers and designers seem to forget that in the workplace, most of that stimulation should come from the task at hand, not the poor lighting, noise, flawed equipment, and information overload that a third of office and a half of factory workers complain about. Even when employees aren't paying attention to them, these background stimuli boost their arousal levels, drain energy from their work, and leave them out of sorts at the end of the day, wearied by the unconscious effort of suppressing continual distractions. Workers who want to improve their environments in order to increase their efficiency aren't asking for the moon: the big items on most lists include quiet, a decent chair, easy access to tools, enough space to maneuver in, and the right to change furnishings around. Despite the obvious benefits to employees and employers both, however, the former are almost never consulted about the design of the places in which they do their jobs.

If too much stimulation stresses an adult, it has more serious consequences for a developing child. "There's a lot to be said for the opposite extreme, typified by a certain kind of nineteenth-century upper-class English child rearing," says Myron Hofer. "The combination of an environment in which there were some books, a lot of experience with nature, and not many people produced a special kind of individual with a rare inner fantasy life that wouldn't be produced in the very different environment of, say, a kibbutz. There has been a lot of debate about whether the very diffuse early environment of these communes, which produced people who were superficially very well adjusted, might have caused a lack in some of the finer discriminations in relationships and inner direction. The most important point to make, however, is not that either end of the spectrum—the kibbutz or the Secret Garden—is the only environment in which to raise children, but that some kids will do well or poorly in either."

America's experimentation with the "open school," one of the most popular architectural developments of the environmentally minded sixties, supports Hofer's assertion. Consisting of a few large unbroken spaces, the open school is filled with interesting things for groups of students to do, see, and hear. The premise reflected by its free-wheeling design, which provides, say, only four corners per 100 children versus the usual 16, is that children learn best in an exciting, hands-on environment that gives them considerable choice about their activities. Both because they seemed so much livelier than the dull old "egg-carton" model and because they are cheaper to construct, many open schools were built. According to surveys, however, the average achievement of the kids who attend them is either the same or worse than that of children in old-fashioned settings. While some students prosper, others find the informal atmosphere and abundant stimuli too distracting. Not surprisingly, noise is the stickiest problem; even though many teachers use bookcases and partitions to reduce visual stimulation—thereby cutting down the amount of truly open space—there's little they can do about the background hum.

The importance of maintaining the right degree of stimulation for each individual tends to be the first thing scientists who concentrate on development bring up in conversation. Even the fetus is busy absorbing input from its environment, albeit in a way that is almost impossible for us to understand. "The fetus isn't having inner experiences—it's not self-aware—but doing things, taking in stimuli," says Hofer. "There's no 'me' in there that, for example, knows its mother's voice. The fetus simply behaves more toward that voice, wanting to keep in proximity to it. It's adjusting its template so that when it's born, it will be able to discriminate that voice from others." Propelled by the genes, this interaction between fetus and environment helps evolve not only a behavioral template but also the structure of the brain itself. Just how remains unclear, but a kind of use-it-or-lose-it principle seems to apply. If the right neural activity is prompted from the fetus's internal or external milieu at the right time, certain neurological connections are forged; rodents and primates raised in complex environments even develop heavier, better brains than those reared in duller settings. If the right stimuli aren't forthcoming, however, those potential links fail to develop or atrophy instead.

The recognition that stimulation is crucial to development has inspired many misguided attempts to soup up that process and, with the help of environmental devices, somehow create superbabies. "The mother herself is such a rich sensory stimulus that a baby doesn't need a lot of equipment," says Anneliese Korner. "These gadgets parents buy with the idea that they can accelerate development or substitute for the mother's time are ineffective, and the overstimulation they can supply can be a problem. Even a mobile may be wrong at a certain point. Parents should watch for cues, such as crying, that something they offer is too much."

According to Gerald Turkewitz, the very young baby experiences the world in just such simple terms of too much, too little, or all right. "For the first two months, the infant responds to the environment as if he were a pure mathematician, not asking what a thing is, but how much stimulation it provides," he says. "His world, which is organized in terms of amount, is extraordinarily different from ours. To him, two things that provide the same amount of excitation—touch, smell, whatever—are equal. For example, in one study that measured babies' heart-rate responses, they reacted almost the same to a bright or dim light as to a loud or soft noise—their physiologies said the two things were the same to them. That's why a baby who is too cold will stop crying for a bit if you rock or feed him. He doesn't care whether he's cold or hungry, just that the amount of unpleasant stimulation is great. We can respond to the environment like mathematicians, too, but we have to work at it. That's no longer the way we see the world."

Even after their first, numerical world view changes, beginning at about six months, children remain highly sensitive to levels of stimulation, and that vulnerability is heightened by a lack of control over their environments. In the 1960s, when social scientists turned their attention to the ghetto, they theorized that children there suffered from a kind of sensory deprivation. "Anyone who thinks of the inner city in those terms has never lived there," says Peter Suedfeld. "What people really mean when they say children in the ghetto are understimulated is that they don't go to enough museums and libraries. I lived in Harlem for a few years in childhood, and one thing I *wasn't* was understimulated." Up-to-date research bears him out.

Stimulation in the form of responsive toys, lots of books, lively decor, and freedom to explore the world does foster good development in early life. However, what is good in short doses on a focused level—say, a toy that "talks" when a button is pushed—can be harmful as a chronic background, such as a droning TV. The economics of poverty decree that the homes of poor kids are likelier to be more crowded, therefore usually noisier and more cluttered, and to lack the privacy and "stimulus shelters" children need by their first birthday.

"In the seventies, a fascinating study of how the urban poor cope with too many people in too little space was suppressed for political reasons," says Ralph Taylor, an environmental psychologist who is a professor of criminal justice and an associate dean at Temple University. "The researcher, Al Scheflen, videotaped people in their homes over time. The films showed that when space is tight, people have to make complex adjustments to find places that offer what they need, from privacy for a phone call to space for a game of hide-and-seek. In a typical sequence, the kids would be in the kitchen doing their homework. When the mother comes in to make dinner, they pick up and disperse to the bedrooms or the living room, where someone else might be blaring the TV. When the father comes home, he can't find a quiet place to have a beer, so he goes out to a bar, explaining that there's 'no place' for him. Different individuals cope with this challenge with varying degrees of success."

In a classic study of the insidious effects of chronic environmental overstimulation on children, psychologist Sheldon Cohen studied the impact of traffic noise on young kids who lived in an apartment tower built over a busy New York highway. The results showed that those who lived on lower floors had worse hearing and reading skills than those on quieter upper floors. Just as the strange noises, shadows, and smells that alarm us on our first night in a new house seem to fade over the next few days, the children in the noisy apartments "got used to" the racket. Their ability to do so depends on a dual reaction involving physiological habituation—neural receptors fire less often after exposure to a stimulus—and psychological adaptation—the tendency to dismiss something we've already considered. If we weren't able to acclimate, we'd be in big trouble, jumping from our chairs at every honking horn or siren. But we pay a price for this adjustment; it is almost as if we spend energy meant for better things on buffering distractions. In the process of tuning out the noise, it seems that the children living over the highway also screened out audio cues required to discriminate sounds, which is important when learning to read. The consequences of this adaptation could be particularly serious for babies, not just because they are learning to talk, but because their neuromuscular development is finely synchronized with their parents' speech.

The invisible pall it casts over learning makes noise a major problem in schools, where levels sometimes surpass federal safety standards.

Near airports, busy roads, and train lines, students' achievement dips as the din rises. One study showed not only that kids on the noisy side of a school had lower reading scores, but also that 11 percent of teaching time was wasted there. Another revealed that compared with other students, those who lived near the Los Angeles airport had higher blood pressure and were less able to solve problems and handle distraction; on the other hand, the same children's math ability was unimpaired. As studies of the effects of noise in the workplace also show, environmental stressors can have very specific and debilitating effects.

If all the research on the best environments in which to raise and educate children could be boiled down to three words, they would be Small Is Beautiful. Intimate surroundings, a low student-to-teacher ratio, neighborhood rather than regional facilities—these are the kinds of nongimmicky, less-is-more environmental influences that year after year have proved to foster both academic and social learning. "We know exactly what's needed to increase the effectiveness of inner city schools by two hundred per cent," observed Harold Proshansky, a father of environmental psychology, shortly before his death. "Dedicated teachers, support services such as meals for kids who need them, and small classes. My wife has worked in what are called alternative public schools in Harlem for twenty years, and they're effective because they don't enroll more than two hundred children."

Reversing their earlier opinion, developmental scientists now think that the real environmental problem faced by inner-city kids is a chronic sensory overload that drains energy from their proper pursuits. Study after study shows that intensive background stimulation hinders development, and that kids' adaptation to it is ephemeral at best. This body of research makes a vital point about the effects of environment on behavior in general: adaptation to a stressor is not the harmless global amelioration that even scientists once assumed it to be. We pay a price for it, often a very specific one. After extensive studies of people's reactions to almost any kind of setting imaginable, this is the main thing that strikes Peter Suedfeld: "To assess an environment's effects on behavior, you have to consider not only the level of stimulation it provides, but also what that means to the person experiencing it."

11

DIFFERENT
PEOPLE,
DIFFERENT
WORLDS

J UST AS places can be assessed in terms of their potential to stimulate, individuals can be described in terms of their capacity to be aroused. In the warm fall sunshine of the National Institutes of Health "farm" in rural Maryland, with research psychologist Stephen Suomi's young primates leaving their quarters for their recreation area, it is hard even for an observer aware of the dangers of anthropomorphizing not to think of kids cutting loose in the schoolyard at recess. The animals' playground is filled with the same hanging tires and monkey bars that make toddlers near the park squirm from their strollers. And like little kids, the young monkeys explore and exploit their turf in what Suomi has learned is a very predictable fashion. "The boldest go first," he says. "The timid may not enter the yard at all, or just go halfway. But however adventurous, they all charge back to their home cage several times in the course of an hour, as if making sure it was still there. Once they touch base, they go out again, always in the same order and always some more often than others."

While the monkeys illustrate Suomi's remarks at top speed, he explains the powerful attraction that draws even the daredevils back home. "A secure base is a safe haven to explore from and return to when the world feels dicey," he says. "There's no question that such a place can change physiology, bringing down cortisol levels and reducing arousal in the face of novelty or stress. If you suddenly eliminated access to their bases, the boldest monkeys would get excited, then they'd start adjusting. The more timid ones would have a much harder time calming down. Because they seem to have a perceptual trigger that causes them to find more things worrisome and generally worth paying attention to, they depend more on familiar settings. You can't help picking up the concept that the familiar place—home—generates a secure feeling, and the novel one—playground—a sensation of potential danger, and that individuals differ in how great they want that mix to be. To interpret an environment, you must not only describe it but also the reactions of different individuals with different genetic and experiential backgrounds to it."

In the course of studying why members of what seems like a homogeneous group react differently to the challenges posed by everyday environments, Suomi has identified two types of individuals: a high-strung sort he calls "uptight," and a relaxed sort he labels "laid-back." He is often able to identify individuals of both types at birth, and to predict the tenor of their future responses to the world. When faced with a novel or stressful situation, Suomi has found, the uptight respond with a high degree of physiological and behavioral arousal, while the laid-back stay calmer, inclined to react with curiosity rather than fear. "Most of the time, you can't tell the timid from the bolder monkeys," he says. "Certain settings, however, such as the playground, have very different consequences for the two groups. One feels challenged to explore, while the other holds back, looks fearful, and acts anxious for longer periods." Rather than "laid-back" and "uptight," researchers who study human temperament use the terms introvert and extrovert, inhibited and uninhibited, or anxious and bold to express similar orientations. Whatever set of labels they prefer, scientists increasingly regard behavior as the product of an individual's effort to match his physiological and psychological makeup with settings that can help him maintain an optimum level of arousal.

Frank Farley may be the president of the American Psychological Association, but he's also a *bon vivant* and long-distance balloonist whose license plates read DR THRIL. In the course of his research at the University of Wisconsin, he has coined his own terms for the basic ways in which certain types of people engage the world. He labels the sort of unflappable individual that Suomi calls laid-back as a "Type-T" (for thrill-seeking) personality, and the uptight sort as "Type-t." "Think for a minute about how different individuals deal with various degrees of certainty, both mental and physical," he says. "Those I call 'Type-T'—for thrill-seeking—are so highly engaged with risk, change, intensity, complexity, and novelty that the pursuit of those things is almost their hallmark. At the other end of the spectrum, Type-t's try to avoid those same things, making a way of life out of safety, predictability, moderation, simplicity, and familiarity. Although both the thrill seekers and the law-and-order group are clearly adaptive for the species, most people—I call them the invulnerables or survivors—are somewhere in the middle."

A canny observer in a hospital nursery could pretty accurately sort the laid-back T babies who are inclined to grow up calm under fire from the fussy, uptight t's, apt to remain easily stirred up. Both these orientations at least partly spring from variations on the human nervous system. A long-term British study of 101 teenaged boys, for example, suggests that the behavioral fireworks of the T-type are signs of his effort to balance a chronic internal calm with external excitement. After measuring their subjects' resting heart rate, skin conductance, and brain-wave activity, the researchers found that the boys whose levels of physiological arousal tended to be low were likelier to be involved in criminal activity ten years later. "In broad strokes, you might say that the T's nervous system is underarousable, which is suggested by their EEGs, while the t's seems overarousable," says Farley. "Our optimum state is the midpoint between boredom and anxiety. To get there, the t habitually avoids stimulation, while the T chronically seeks it."

Accounts of seemingly nerveless thrill seekers abound in narratives of the Civil War, fought on both sides by many officers who participated in the Mexican and Indian conflicts on the frontier. In fact, descriptions of many of its heroes recall those of the lean, laconic cowboy types with reptilian pulse rates who really won the West.

The very prototype of the T-type is Colonel John Singleton Mosby, the Confederate guerrilla who led the Union army such a merry chase within sight of the Capitol dome—he once kidnapped a Northern general from his camp bed—that after the war he was befriended by an admiring Ulysses S. Grant, himself a low-key sort legendary for his utter calm under fire. "He is a different man entirely from what I had supposed," his former adversary wrote of Mosby. "He is slender, not tall, wiry, and looks as if he could endure any amount of physical exercise." Other contemporary accounts strike the same wondering note that the perpetrator of such heroic exploits might be so unassuming in mien. "He is slender, gaunt, and active in figure," wrote J. E. Cook in *Wearing the Gray*. "A plain soldier, low and slight in stature, ready to talk, to laugh, to ride, to oblige you in any way—such was Mosby in outward appearance. Nature had given no sign but the restless, roving, flashing eye, that there was much worth considering beneath . . . [but] the commonplace exterior of the partisan concealed one of the most active, daring, and penetrating minds of an epoch fruitful in such."

These descriptions might have been written about mountaineer-scientist Peter Hackett, who not surprisingly looks for just such kindred T-types when assembling teams for the rigors of a high-altitude expedition. "Given two options, one of which has a definite outcome and one of which doesn't, some people will choose the one that they aren't certain about," he says. "To them, an iffy situation that might test their skills is somehow more rewarding. That kind of person, who could just be more naturally curious, is often identified as a risk taker." The criteria he uses to separate the T's from the t's sound as if they had come straight out of Farley's and Suomi's data, or from the Soviet studies on the unflappable types who do well in Siberia. "I look for people with a marked flexibility and ability to tolerate upsets—more of a laid-back personality," he says. "They might be intolerant of crummy food or service in a restaurant down here, but up there they can wait out a storm on a rock ledge for four or five days. The high-altitude environment poses a lot of stresses—weather, avalanches, crevasses, accidents, illness—and a climber's greatest attribute is the ability to handle prolonged discomfort."

Even in conventional environments such as homes and parks, Farley has found the T prefers the more complex, variable, active versions. "On a microchip assembly line, a T is a disaster," he says. "In fact, that image pretty much illustrates my theory of stress. If you read those magazine stories about 'The Ten Most Stressful Jobs,' the list always includes cops, but my guess is that a lot of them are T's who love what they do. There are very few absolutes in stress. To a cop, his work isn't stressful—it's thrilling. Some T's concentrate on physical excitement, while others, such as entrepreneurs, go for the mental sort. Some people go for both."

The U.S. population comprises more than its fair share of all three types of T's. "America began as a T nation," says Farley. "Our origins go back to risk-taking explorers, adventurers, and immigrants. Even today, the T's still comprise about twenty to thirty percent of the population, as opposed to the t's ten to fifteen percent. Those proportions would be quite different in, say, Switzerland. Talking about national differences isn't very popular right now in academe, but it's a concept used in the real world all the time. One legitimate way to study why a large proportion of a country's people are inclined to behave in certain ways would be to look at immigration records. For example, very few people relocate to Sweden or Japan, but many move here and to Australia. Those patterns have to have implications for both the nations that receive immigrants and lose emigrants, who tend to be T-types."

According to Farley, America is a T country not just because of its gene pool, but because its culture encourages enterprise, independence, elbow room, and a free-market economy. "Capitalism is a T invention," he says. "Something that worries me is that literally and figuratively, we're not providing much range for our explorers to roam in anymore. The writers of the Constitution had a vision of a nation based on the tension between T and t, between experimenters and conservators. But as America has gotten older, the legislators have piled up more and more laws, because that's what t's do. A hundred years ago we were opening frontiers, and now we're afraid to eat an apple. At least the next century will belong to T's, because our species is leaving the earth for space."

In most of America, says Farley, the T's and t's combine pretty much in the right proportions with the mid-range majority of T/t "invulnerables" to produce a society that allows for occasional bursts of wildness and risk taking within an overall framework of security and order. But in some places, a higher than usual proportion of T's creates an almost palpable atmosphere of excitement. These days, most of this explosive chemistry between extreme settings and temperaments occurs in big cities, but not so long ago, it permeated the frontier. These disparate urban-rural attractions pull at the many hard-driving city folk who head for the hills each weekend, says Farley, observing that what might seem like a reasonable compromise—the suburbs—isn't even an option for them. "People involved in city-country polarities are probably T's who are drawn to both extremes for different reasons," he says. "My guess is that the tranquil environment wouldn't remain appealing to them over time. It's great as a contrast, though, particularly because many city people have active mental lives and get a lot of their stimulation internally."

Although the American frontier has shrunk, stretches of that kind of wildness still exist in the Far North, affording us a glimpse of the interaction between new country and the pioneers it attracts that forged the nation. "There is no doubt that Alaska draws a risk-taking personality who likes the idea of being able to do pretty much what he wants, with minimal interference from others, including the government," says Fairbanks psychiatrist Irvin Rothrock. "It's like the West was a hundred years ago." It makes perfect sense to Farley that Alaska would be the last continental bastion of T's, as well as grizzlies and gray wolves. "The challenge of the physical environment is so powerful for them," he says. "Uncharted territory is their true venue."

Overwhelmingly outnumbered by the young male thrill seekers attracted to the state, Alaskan women are sometimes envied by sisters in the urban "girl ghettoes" of the Lower Forty-eight. However, as one female resident of Homer puts it, "Up here, the odds are good, but the goods are odd." According to Rothrock, Alaska doesn't really have more mental illness overall, but like certain other places, an under- and overrepresentation of particular disorders. "I trained at the Menninger Foundation in Topeka, and depression is a real common problem in Kansas. The traits of people who tend to get major depression—the preoccupation with detail, the high degree of responsibility, the compulsiveness—fit very well in that kind of farming society. Kansans are serious people who work hard and don't fool around with frivolous things. I very rarely saw mania there. When I was on the faculty of the university, if we got a manic patient on the ward, we'd tell all the residents and medical students to drop by and talk to him, because that was something they wouldn't see too often. A manic personality looks around Kansas for a couple of days and says, 'Hey, this is not for me.' But here in Fairbanks, in our little eight-bed psychiatric ward at the hospital, we have a couple of manic patients at a time. There's a much higher percentage of bipolar people who tend to get manic here than in the Lower Forty-eight."

In trying to relate an individual's thrill-seeking propensity to various disorders, Farley has found some evidence that T-types tend to have a bipolar streak in their characters. Rothrock agrees that the entrepreneurial, free-spirited, larger-than-life nature of Alaska appeals to the manic type of personality for the same reason New York and Washington do. "Los Angeles and the northern frontier are physically different, but they provide the same opportunity to wheel and deal and operate, which that kind of person needs to feel comfortable. The better-organized, mildly affected individuals who are just hypomanic are extremely energetic, enthusiastic, and productive. They just go twenty-four hours a day. You can't keep them out of the leadership stratum. It's an interesting side of Alaska. As long as they don't get too ill, people with bipolar tendencies get along very well up here."

If mountaineers, stockbrokers, and plenty of Alaskans are at one end of the arousal spectrum, the Type-t Prousts among us are so revved up from within that they try to keep excitement from the external world to a minimum, eschewing rock ledges and the trading floor for claustral settings. Emily Dickinson spent almost all her life in Amherst, Massachusetts, and ended it as a recluse in her father's house, yet her poems prove that even in such controlled, subdued settings, she was deluged with stimuli, almost unbearably so. "There's a certain Slant of light,/Winter Afternoons—/That oppresses like the Heft,/Of Cathedral Tunes," she wrote. And "What Fortitude the Soul contains,/That it can so endure/The accent of a coming Foot—/The opening of a Door—." Her more robust countrymen have generally tended to disparage such a high degree of sensibility, yet Type-t's are well represented in the gene pool, suggesting their traits have also come in handy during evolution.

However lightly t's might seem to tread through the world, their inward focus gives them an inflexible attitude about it that is sometimes more productive than that of the extroverted, adaptable T. Unaccustomed to looking for outside feedback, t's are far more inclined to tailor their surroundings to fit their mind-sets than the other way around. If bold T's led the way through the wilderness, pushing the frontier to the Pacific—one informal genetic explanation of why Californians are so adventurous—it is the sensitive yet stubborn t's who stayed in Kansas to pummel it into civilization. "In a dangerous environment, it's useful to be cautious, whether about avoiding predators or trying new foods," says Suomi. "Being the uptight type has social benefits, too. Because shy, anxious children spend more time watching than doing, for example, they're likelier to become the scientists, artists, philosophers. They might not be able to short-circuit their physiological arousal reaction entirely. But with some experience in managing their environments, they can get a handle on it, and perhaps even learn to enjoy a beating heart and an adrenaline rush."

That someone born high-strung might learn to enjoy acting laidback shows that while genes may be a powerful influence on our basic approach to stimulation from the world, they are not the only one. Our experience, particularly the early sort recorded by a still-developing brain, powerfully shapes that orientation as well. Timothy Leary, an expert on temperament before he dropped out of academia, is prone to ask people about the circumstances of their lives when they were seventeen; he has found that our basic attitudes about ourselves and the world tend to be meshed into the rapidly setting concrete of our nervous systems around that age, and usually change very little thereafter, which makes a youthful biography a good index of adult personality. This observation perhaps explains not only why life continues to resemble high school, but also our conservative streak when it comes to choosing environments. "It's hard to change preferences for certain places that were imprinted during sensitive periods, when our brains were enjoying a growth spurt," says Suomi. "If we loved the country and hated the city then, we probably still do. We might very well be so nostalgic for our youth and its haunts because of what our brains were up to at that time."

This formative power of experience, especially the early sort, ensures that inborn tendencies needn't be life sentences. Much of what we assume is "instinctive" behavior has actually been learned in early life. A fear of snakes seems inborn, for example, but while monkeys who grow up in the wild have it, those raised in a lab do not, because they haven't been exposed to the frightened reactions of other monkeys. Of all early experience, the relationship between mother and infant is the most important influence on our behavior, and its dynamics powerfully shape the baby's current and later responses to the larger world. In fact, Suomi has found that the single most effective way to modulate an infant monkey's inborn traits is through a propitious switch in maternal environment. "Even though we can deliberately breed timid or bold individuals, we can substantially modify the anxious trait by manipulating a monkey's early setting to reduce its arousal," he says. "While a laid-back baby isn't necessarily influenced that much by an anxious foster mother, an anxious infant adopted by an unusually protective, reliable mother will learn to handle stress as well or better than a genetically bold baby. With her as a secure base, he can learn to handle new situations, watch out for trouble, and even become one of the group's most dominant members. This same secure-base dynamic applies to physical settings at least until puberty, and possibly later in life."

Noticing that rat pups became especially hyperactive when deprived of their mother in an unfamiliar physical environment, Myron Hofer decided to find out why. Experiments showed that a low dose of reserpine, a drug that prevents the storage and transportation of agitating neurochemicals called catecholamines, completely eliminated the upset reaction. It seems that, like the drug, a mother regulates her offspring's neurochemical responses to the world, particularly its new, more stressful parts. Without her buffering presence, the infant's catecholamines rise, producing hyperactivity. Further studies showed that while her heat, smell, and particularly her tactile stimulation each separately helped prevent agitation, together they provided a stronger, longer-lasting calming influence. Social scientists have long observed that babies separated from their mothers too often or too long are apt to be upset by unfamiliar situations, even later in life. Hofer's study implies that this anxiety may derive from the combination of a learned response and an environmentally mediated neurochemical predisposition as well. "From an evolutionary point of view," he says, "the mother's regulation of the infant's physiology and behavior has offered us mammals a great adaptive advantage. It allows a rapid change in the biological characteristics of a species from one generation to the next, according to the changing requirements of the environment. For example, a mother can preadapt her infant to a cold climate by providing more nursing and nesting. Maternal regulation is a nongenetic way of passing characteristics on to the young. It allows the experience of the parent to mold the traits of the offspring."

Rather than thinking about our environmental interactions in terms of our individual neurological wiring, as Suomi and Farley do, other scientists approach the "person/environment fit" from the group perspective of people who share special needs. Ramps leading into public buildings and rest rooms designed to accommodate wheelchairs are constant reminders of the difficulties faced by the handicapped, but they are hardly the only environmentally underprivileged group. During the first dozen or so years of life, for example, children constitute a much larger disadvantaged minority. Usually assigned the oddest, smallest, least private spaces, they must sandwich their needs around those of adults, at least when their elders are at home. Most of the time they are not, yet very little attention is paid to modifying domestic settings to make life more secure for unsupervised older kids or to the design of day-care facilities for the younger ones. In the larger world, children are often too small, weak, inexperienced, and unmonied to function competently, as their high rate of accidents proves. Because of children's mental as well as physical immaturity, places that seem perfectly safe to adults can be very hazardous for them. One of the worst scenarios for accidents, for example, is a setting in which kids suddenly confront fast, heavy traffic, as when a young ball player or cyclist suddenly turns a corner shielded by shrubbery onto a busy thoroughfare. Yet the adult community is rarely aware of such trouble spots until a tragedy occurs. "Until they're about twelve, kids just don't react rationally to cars—they need us to physically protect them," says Craig Zimring, a professor of psychology at Georgia Institute of Technology. "They're involved in even more accidents as car passengers, yet auto designers focus on adults when they make safety devices."

It is sometimes said that the two sexes "live in different worlds," and there is some evidence that is so. The traditional explanation for the tendency of the female to have greater acuity in every sense but sight is that she needs keen auditory, olfactory, and tactile capacities to keep her offspring safe and well fed, while the male requires keen eyesight and spatial perception for hunting and building. "The woman certainly lives in a quite different internal environment than the man, whose brain is bathed by different hormones," says Paul MacLean. "She certainly doesn't visualize and go for pornography the way the man does. And when the baby cries at night, although the mother thinks her husband is just pretending to be asleep, I can honestly believe he's deaf to the sound. I think there's probably something to the idea that the woman is especially attuned to the baby's tiniest noise. If the horse has grown sensitive to the rider's knee because of a long history of being ridden, why wouldn't the mother's ears react to the baby's cry in a way that no one else's would?"

While boys and girls may start life inclined to perceive the world differently in some respects, the cultural discrepancies in their experience of it have more profound effects. For example, little girls often don't map their neighborhoods as well as boys. However, at least part of the reason girls are not as adept as boys at such tasks is because their exploration of the physical environment is far more restricted; they are not allowed to go out as often or as far. When girls' curtailed experience is overlooked and differences in their performance are solely attributed to biology, the traces of a familiar vicious circle surface in many cultures.

Like kids, the huge group at the other end of the spectrum of years often suffers from a poor fit with and lack of control over its environments. Even the 4 to 5 percent of old people who must live in institutions do better when their setting is organized around what they can do rather than what they can't. Power over superficially small things, from the care of a plant to doors with knobs that can be turned by arthritic hands, as well as access to cooking, laundry, and recreational facilities, ensures a much better quality of life. Residents who have their own rooms are spared many environmental threats to well-being, such as crowding, noise, and especially loss of autonomy; not surprisingly, they are much more sociable and active.

In a chilling contrast, one study showed that those who must share a room spend *half* their time lying in bed, whether they are asleep or awake.

Anyone who has kept a vigil with a friend in a hospital has experienced something like the kind of chronic environmental cold shoulder endured by the handicapped, the aged, and children. From the moment visitors enter the fancy, spacious lobby devoid of friendly amenities such as seats and clear maps, we are made to feel invisible, if not actually unwanted. Already worried and intimidated, we must trace through what often amounts to a maze to reach the patient. Once there, we may have to spend considerable time in a stark waiting area, usually lined with uncomfortable chairs that provide little rest and discourage communication. We need privacy and peace, but are subjected to scrutiny, noise, and bright light. Simple amenities like coatracks, water fountains, bathrooms, and phones are usually inadequate. And if we are provided with little information about what goes on in the hospital, we learn far less about services in the neighborhood, and must puzzle out the locations of transportation, restaurants, and shops on our own. We need a little bit of home away from home but feel homeless instead—a sure sign of bad *feng shui*, or a poor person-environment fit.

In such inhospitable places, we are less likely to enjoy the optimum sort of experience that Mihaly Csikszentmihalyi calls "flow." To measure the supposedly unquantifiable quality of life, he randomly pages a subject by beeper eight times per day for a week; at the signal, the subject writes down what he is doing, from sweeping the kitchen to writing a poem, and also notes how he feels about himself at that moment. After analyzing some 25,000 of these highly individualized "experience sampling" reports gathered over twenty-five years, Csikszentmihalyi says that "despite the scientific prejudice, subjective experience is one of the most objective things we can deal with." He has found that far from being unpredictable, our responses to life vary closely with our ability to meet the challenges it poses, and that our environments play an important role in these interactions.

The most frequently cited examples of flow concern the high moments of musicians and artists, bungee-cord jumpers, and white-water canoeists. But a European psychiatrist familiar with Csikszentmi-halyi's work decided to see if flow research could help a schizophrenic patient who had been hospitalized for fifteen years. When the quality of this woman's experience was monitored, the record showed that she had been happy twice during that week. On both occasions, she had been cutting her nails. Based on information of a sort that he would not normally have acquired, the doctor suggested that the woman receive training to become a manicurist. The patient loved this work, and was eventually released from the institution to pursue her career in the community, requiring only a weekly visit from a nurse. Whether it is a manicure salon or a physics laboratory, a kitchen or a cockpit, most of us similarly depend on places to provide the external framework that helps us structure our inner lives and bring about flow.

"If the average person goes to Yosemite and sees a rock wall, there's nothing inherently exciting about it," says Csikszentmihalyi. "But if the visitor is a climber, the rock means a certain action potential. To him, it's the most important part of the environment—perhaps of his life. In the same way, to most people the sea is just a big puddle of wetness, but to a sailor or a fisherman, it's an entirely different thing, because of what that particular person can do with it. People who experience flow while engaged in sports such as hang gliding have a bad reputation for wanting to take risks, but I find they're actually control freaks who enjoy proving that the environment won't get the best of them.

"When we're in flow, whether while playing the violin or climbing a mountain, our actions merge with our awareness. We stop being spectators of our own experience, which eliminates that ruminative self-consciousness that's such a burden. We feel a sense of oneness with something larger than the self, whether it's a musical tradition or nature or a deity. Because we're concentrating on the present, our activity dictates our experience of time rather than the clock. This intense focus also means we forget our daily problems, which even six- and seven-year-olds say is important to them. People complain about children's absorption in video games, but if we studied how these entertainments produce flow, we could make better educational tools. After all, according to Plato, the task of educators is to teach the young to take pleasure in the right things."

While analyzing the daily experiences of a group of high school

students, Csikszentmihalyi found that like their elders, his young
subjects felt best when they were hard at work on a task in which
they were proficient, and worst when stuck with an activity for which
they had little aptitude. For some, flow occurred in a classroom, and
for others, on the basketball court; as Csikszentmihalyi says, "Neither
good nor bad per se, flow is just energy." Like the students' reactions
to the day's challenges, ours can also be plotted on a spectrum of
arousal. When we move beyond flow to the point at which our abilities
dwarf the demands made upon them, we slide into a complacent
state Csikszentmihalyi calls control. As the stimulation level continues to fall, we inch into boredom, relaxation, and finally, apathy.
Beyond this point, marked by poor skills matched by negligible challenge, we head toward the least comfortable end of the scale; as our
abilities shrink and the demands upon them grow, we pass through
worry into anxiety, which signals that we're in way over our heads.
Not surprisingly, Csikszentmihalyi has found that self-esteem and
mood are poorest at this point. "We spend more time at the understimulating end of the scale because we feel so bad at the high end
that we'll do almost anything to reduce our arousal," he says. "Most
people experience an almost three-to-one ratio of boredom and apathy to anxiety. That's partly because when we try to relax, instead
of ending up at an optimum level of stimulation, we overshoot. We
wind up in a state of apathy, which is what most people report when
watching TV. After a stressful day at work, they pass a dull night in
front of the television."

Because access to the right settings, from basketball court to classroom, workshop to office, is so important to attaining our peak
experiences, many people report that Sunday is the low point of the
week. In fact, what psychologists call "the Sunday neurosis" can be
defined as the antithesis of flow. This dreary state sets in when,
deprived of access to the computer, the rock wall, or the nightclub,
we're challenged only by the newspaper. "One of the worst things
that can happen to most people is involuntarily having nothing to
do but dwell on the self, which is often the case on Sunday between
ten A.M. and noon," says Csikszentmihalyi. "The mind immediately
turns inward and starts reviewing what's wrong. By Sunday afternoon,
most people feel better, because by then an activity has usually
emerged."

The kinds of activities they offer make some places more popular in general than others. "People are usually happiest in settings in which they're relieved of others' demands and in control of their own actions," Csikszentmihalyi says. "We tend to like parks because they offer lots of options, from climbing a tree to taking a nap. Just the possibility of walking in any direction that such a place offers gives a sense of freedom, accentuated by nature, that built environments lack. For some of the same reasons, by and large we very much enjoy being in cars, where we command our own little portable worlds and can listen to music without any responsibilities to deal with."

Along with our preference for wide-open spaces, at least in a figurative sense, Csikszentmihalyi has found that we're generally happiest in public settings with other people around, whether it's outdoors at a park or beach or in a theater or a restaurant. For teenagers, that's particularly the case. Far and away the most positive place for them is a park, because they're usually there with other teens, which always improves their mood, and there's no adult supervision, which increases their sense of independence. "You can't just think in terms of places, but of what they require and with whom you share them," he says. "Even within a single school, there are enormous differences in what different spaces offer and demand. That's why the cafeteria and student center are the most popular spots and the classrooms the least, while the halls rate somewhere in between."

Just as most of us feel happiest in parks, cafés, and other sociable, carefree spots, we're apt to feel more constrained in the workplace, where our responsibilities and others' expectations are many and opportunities for self-determination can be few. Most of us at least occasionally flow between nine and five, but estimates vary from over half the time for managerial types to a third of the time for those in blue-collar positions. One reason why some kinds of work produce more peak moments than others is that in order to flow, we must have clear goals and fast feedback. That's why so-called pink-collar workers, such as clerks and secretaries, who are at others' beck and call, report a lot of work-related stress, while surgeons are right up there with musicians and bungee-cord jumpers at the zenith of flow. According to Csikszentmihalyi, "Some doctors say it's as addictive as heroin."

At home, the two sexes are inclined to seek different places to help create flow. "Men often consider the basement to be the best part of the house, while women favor the bathroom," says Csikszentmihalyi. "To a man, the basement is a retreat, a place for hobbies, a workshop, a bar, or a game room, but to a woman, it means loading the washer or a problem with the boiler. In the bathroom, surrounded by the technology that helps her establish her identity, she feels safe. These sorts of gender differences apply to many parts of the house. For example, while the kitchen is usually more important to women, it represents obligation as well as pleasure to them. It's only the setting for flow when they cook by choice. Both men and women tend to like the bedroom, but their good moods there may not be a reflection on the room per se, but because that flexible setting allows them to be alone, read, or make love."

Unless we're among Csikszentmihalyi's subjects, most of us face a formidable obstacle to living a more enjoyable life. Not unlike the schizophrenic woman before she became a manicurist, we lack detailed information about when and where we feel best. Partly as a result, we resign ourselves to our dreary offices and inconvenient kitchens, hanging on until we can get to the beloved quiet study, the seaside cottage, or the bowling alley for some "quality time." Taking a cue from flow research, one urban workaholic was able to improve his life by analyzing one of his drab interim environments and making some changes in it. For years this man had complained that he didn't get to his country place often enough because of the long, dull drive. One of his other gripes was that he was so busy keeping up with professional journals that he never had time to read for pleasure. Finally inspired to put two and two together, he decided to listen to books recorded on tape when commuting. A boring waste of time became an opportunity for flow, and his neglected house was transformed into a beloved refuge. "Before you can use the environment to improve your experience, you need a lot of data about how you live so you can do some fine micromanagement," says Csikszentmihalyi. "It's often the small parts of life that are the hardest to manage. In the course of doing a study for a big Japanese auto maker on how cars can be made more enjoyable, for example, we talked to many people who drive a lot and got some excellent, concrete ideas, but we really had to focus in to get them. The first step is sitting down to figure out what's going on, one thing at a time."

When the staff and residents of one retirement home did just that, they were surprised by what they learned. First of all, the old folks were paged by beepers at different points in the day and filled out charts noting what they were doing and how they felt at the time. Their responses immediately made clear that a few things were much more important to almost everyone's happiness than supervisors had thought—food, in particular. One man of ninety-two was instrumental in highlighting the vastly underestimated value of a change of scene. His report showed that he had felt depressed all during the week of the test, except on two occasions when he had been wheeled to the elevator. It turned out that this seemingly inconsequential event was his only exposure to people coming from and going to the larger world. Armed with hard information instead of tired assumptions, the staff got busy changing their clients' environment. Meals became more festive. Instead of pushing all the residents confined to wheelchairs into one room to vegetate in front of the television, they were offered a choice of settings. "Once people had the tools to learn what they hadn't known, the place became much more responsive to individual needs, and much livelier and more interesting," says Csikszentmihalyi. "To think that just watching people coming in and out of elevators can be a source of excitement and pleasure! Without the beeper, however, not even the old man would have realized when he was happiest."

In quite a different setting, Csikszentmihalyi used his happiness technology to figure out and remove the obstacles that interfere with the pleasure of visitors to museums. "Most people aren't sure what they're supposed to do in such places," he says. "They know the museum is a good thing, but it's alienating. They feel intimidated by the guards, ignorant about the art, and mystified by the labels." Once these problems were identified, one museum installed computers in its halls that allowed visitors to call up a painting on a screen, zoom in on details, and find out all about it and the artist, even gossipy details of his personal life. As a result, the number of people who enjoyed an exhibit rose from 23 to 78 percent, and visitors spent three to five times longer looking at the paintings. "Other museums have used simpler ways to get people involved, say, by asking them questions and supplying answers, or even by providing clearer labeling," says Csikszentmihalyi. "Making such simple improvements in an environment can completely change our whole experience."

12

.

FROM THE NEST
TO THE GLOBAL
VILLAGE

As THE earth continues to urbanize at a breathtaking clip, finding the right settings for our individual needs becomes more of a challenge. Mexico City, which already has 20 million residents, offers a chilling preview of a new historical development: the megacity, teeming not with millions but with dozens of millions of people. The effects of this kind of environment on our behavior is a controversial subject that has important implications for the future of all species, not just our own.

Like most old-fashioned bucolic suburbs, the area known as Philadelphia's Main Line has changed a lot over the past twenty years. The traffic that throbs along the narrow roads that connect once rustic little towns is so bad that it can seem worth risking one's life to avoid entrapment in a left-turn lane. Most of the pokey little businesses that had flourished unchanged for decades along Lancaster Avenue, epitomized by a combination hardware–feed store that kept a life-sized model of a horse out front, have been replaced by mini malls of generic boutiques, office complexes, and "townhouses." The old-fashioned farmers' market has been relocated to a renovated movie theater, where the stalls of pink-cheeked, white-capped Mennonite women now sit side by side with those of nouvelle pasta and cappuccino artistes. Once horse country, the Main Line's outer reaches are condo country now.

Although Philadelphia seems to have boiled over, spilling into the little towns and pushing back the perimeters of the country, it also seems to have evaporated. The men of the Main Line once rode the trains for which the area is named to their jobs in town, but now, joined by women, they are just as likely to work in suburban "professional plazas." Serious shopping once meant a trip to center city and Wanamaker's legendary department store; children looked forward all year to their Christmas visit, which included an organ recital accompanied by leaping fountains and an aerial train ride around the toy floor. A few years ago the great emporium was whittled down to a splinter, and suburbanites make their purchases at sprawling local malls rather than venture into what some call Filthadelphia.

The Main Line is still one of the East Coast's finest residential belts, and those visiting for the first time must see a prosperous, well-tended suburb that could be lifted and plugged into any well-off area in the country. Despite the beautiful campuses of Bryn Mawr and Haverford colleges and the well-preserved mansions and gardens, however, those who know it better realize that its Quaker village roots have been strangled by overdevelopment. Neither a real city nor the pastoral suburb it used to be, the Main Line has been "urbanized."

Argumentative by profession, social scientists ominously agree that urbanization will be the most important environmental influence on behavior in the twenty-first century. In 1850, 2 percent of the world's population lived in metropolitan areas of 100,000 or more; by the year 2000, driven primarily by the need to make a living, 40 percent will. In the U.S., the 1990 census shows that like Main Liners, 46 percent of Americans now live near cities, but not in them; as is the case with Philadelphia, growth is negligible there. As urbanized sprawl steadily expands, questions of how we handle and share our spaces, from the few inches that surround us physically to the territories of home, school, the workplace, the community, and the ecosystem, take on a new importance, not only for us but for all living things.

Those who look at the ways in which people handle their various increments of turf from a social perspective insist that after thousands of years of living in communities and cities, our cultural adaptations to such arrangements far outweigh any biological predispositions our species might have. As an evolutionary psychobiologist, the NIMH's Paul MacLean takes another point of view, one based not on the history of thousands, but millions, of years. After a lifetime of studying the biological substrates of behavior, he believes that just as each of us has been shaped by the womb, our species has been influenced by its long incubation in the place that helped forge mammalian ways and finally, human culture.

"It all began in the nest some hundred and eighty million years ago, when the family came into being," he says. "Don't forget that reptiles just lay eggs and go off, leaving their offspring to hatch and grow up on their own. Well, our evolutionary transition up from reptiles to mammals was marked by the development of crucial forms of behavior that depended on the evolutionary growth of the brain's limbic system and have helped ensure that we treat our young very differently."

While investigating the evolution of mammalian social behavior, MacLean studied an area of the limbic system called the thalamocingulate division; although it has no reptilian counterpart, it is very large in Homo sapiens. This structure turned out to be so important to parenting that when it was impaired, only 12 percent of the pups of the altered rats in one experiment survived. Influenced by the functioning of the thalamocingulate division and the environment of the nest, mammals developed three types of behavior that separate us from our scaly forebears. First, says MacLean, came nursing and attentive mothering. "Our mammalian feelings of parental concern and social responsibility began with the nesting female. She was the center of things." Next came the vocalization that helps mother and offspring stay in touch, particularly the "separation call" that was perhaps its earliest and most basic form.

The final behavior that separates mammals from reptiles is play. A young hamster with a dysfunctional thalamocingulate division won't romp with its littermates in the normal way, but acts like a reptile. According to MacLean, play is serious business for a higher species whose infants require protracted care in close quarters. "Nests are pretty tight places, but if you're playing rather than fighting, your siblings' teeth don't feel so sharp. Perhaps that's where mammalian sport originated, way back at the very beginning when peace had to be kept in the nest. Remember the chimpanzee mother who would play with Jane Goodall's small son when he was getting too rough with his little sister?"

MacLean thinks the disparity in scale between our modern megaenvironments and ancient origins in the nest can have serious consequences. "The parental concern for the young that started with the nesting female evolved into our broader feeling of social responsibility and conscience," he says. "Now we must extend that limbic sense of family, born in the nest, to all humanity. The challenge is peace, which is hard to preserve because of crowding at home and in the world. Having evolved in small groups, human beings, like mammals generally, have very few built-in mechanisms for dealing with crowds. Except for the pig, the opossum, and a few others, the mammal has between two and twelve nipples and usually bears a limited number of young. We have few limbic wires for dealing with more people than that. And what do people do on those special occasions, such as fairs and festivals, when we do tend to get along in a crowd? We eat and play games together—the two things we did in the nest. The small community has really been the blood and guts of the U.S.A., the place where character is made, where people are close enough to smell, to get to know, each other. We pretend we know everybody these days by using first names for immediate recognition—we've become just like golden retrievers—but we don't really know people anymore."

Since Malthus, scientists and planners have worried about the impact of overpopulation on food and other resources, and that concern must increase as the earth's community of five billion people doubles over the next forty years. Estimating the risks this growth poses for our physical survival, however, is easier than projecting how more elusive behavioral aspects of well-being will be affected by increased social density—the number of people in a particular area—and crowding, which is our subjective reactions to that situation. The modern inquiry into such questions began with the animal studies of psychologist John Calhoun, a descendant of the vice-president and political philosopher, whose work before his retirement from the NIMH is perhaps best known in lay circles as the inspiration for the popular children's book *Mrs. Frisby and the Rats of NIMH*. Early in his career, Calhoun noted that when species that bear immature young leave the nest to get food, a brain mechanism "blinds" them until they get a certain distance away—say, two hundred feet for a rat—thus preventing the depletion of nearby resources. As he continued to explore the links between animals' heredity and their use of the environment, Calhoun became convinced that "for species that have homesites and are territorial, from the Norway rat to man, it's important to live in groups of a dozen, or act as if they do."

In the 1960s, under the aegis of the NIMH, Calhoun began to build his famous "rat universes." In a typical experiment, four pairs were let loose in a predator-free paradise in which all their physical needs were met. At first things ran smoothly. As their numbers rose, however, the rats were unable to split off and find new territory, as they would in the wild, and their behavior grew perturbed. Most males stopped defending and mating in their particular territories, and most females ceased either bearing or rearing the young. Uncharacteristically, some rats gathered in clumps—Calhoun called them "barflies"—while "Pied Pipers" approached strange objects and even followed people.

Rodents that lived in the more crowded parts of the universe lost more of the essence of what Calhoun called their "mousity." Although the environment's design deliberately made it harder to get to uncrowded areas, the most normal males sought them out; they were rewarded with a one-to-seven ratio to females, who also behaved pretty well in these zones. In contrast, the easily accessible zones were crowded with aberrant males, and their ratio to females climbed to twenty to ten. Some rats passively ignored the rest; a group of large bisexuals that spent much of their time grooming were dubbed "the beautiful ones." Then there were "the probers," who lived in "behavioral sinks," or the most crowded areas, where all possible rat pathology eventually bloomed. Dysfunctions among the males included aberrant sexual behavior, cannibalism, hyperactivity, and withdrawal; the females' behavior was such that 80 to 90 percent of their pups died before weaning. In one study, as the birthrate fell and the mortality of the young rose, the rodents' numbers declined from a peak of 2,200 on day 500 to 1,650 on day 965, and to zero on day 1,660, when the last animal died.

Scientists had traditionally held that a population died out either because of predation or a lack of resources. After monitoring his rat universes, Calhoun concluded that the stress of crowding itself could also cause extinction. Some work by colleagues conducted in the wild pointed in the same direction. While studying an overpopulated Chesapeake deer herd, for example, researchers recorded physiological changes in the animals, such as a swelling of the adrenal glands caused by stress hormones, that reduced fertility until the herd died off. Such studies imply that at least for some animals, crowding can be the ultimate form of overstimulation; forced into too many social interactions, they experience physical and behavioral breakdown. Although it is a controversial opinion, Calhoun believes that animal research on crowding has important implications for people. "We too are biologically adjusted to live in small groups," he says. "Our numbers have increased a thousandfold since we began 'to fill the Earth and multiply,' but our heredity hasn't changed. In order to get things done, we've simply learned how to cooperate socially and stay out of each other's way.

"For forty thousand years, the population has periodically doubled, which has meant more ideas, information, and creativity for our evolving culture. But each time our numbers double, the number of social roles and the effort it takes to maintain satisfaction with life also increase. That's one reason why India's caste system, which serves to reduce the number of social contacts, came about. We've become so overwhelmed by the roles, communication, and information produced by our huge population that we've had to become more impersonal. We've become dependent on machines, such as computers, to help us preserve order and complexity and keep us from disintegrating into chaos. These machines also make life so interesting that many people would rather concentrate on information than on raising children. Eventually, the population level will fall."

Those afraid that like mousity, humanity may also be jeopardized by too many social interactions, should cut down on the number that our increasingly urbanized world demands, says Calhoun. "People whose lives are broken up into lots of short segments and appointments have higher rates of suicide and heart disease because they're just overloaded with contacts. We can't even learn from our experiences unless we have refractory periods to digest them in. We can structure our contacts in ways that can help us be happier, but the best most of us can hope for is to have satisfactory social encounters about half the time. In general, the best approach is to try to live your life at any given moment as if you were part of a group of twelve."

Calhoun's animal research on crowding and territoriality became an academic *cause célèbre* of the 1960s partly because it put people in mind of the urban social problems, from fractured families to crime to sexual excess, that were then beginning to attract a lot of attention. Because of the tiny island's huge—and by implication, largely unbalanced—population, professors blithely declared Manhattan to be a behavioral sink.

After this initial burst of enthusiasm for equating big cities and rat universes, however, scientists have grown far more cautious in extrapolating theories about people and their spaces based on animal research, particularly that conducted in manufactured settings. The points of view on personal space, territoriality, high social density, and crowding expressed in Ralph Taylor's book *Human Territorial Functioning* differ markedly from those of biologically minded scientists such as MacLean and Calhoun. "When discussing these issues, I am *very* uncomfortable with sliding from other species to people," says the environmental psychologist. "The kind of biological reductionism espoused by Robert Ardrey and Conrad Lorenz ignores the cultural evolution that also takes place along Darwinian lines. 'Are human beings *meant* to live here or there?' is a good question, but we can't really answer it. There's no way to prove that we are or aren't wired like rodents or primates in terms of how we respond to crowds, because we've become accustomed to big groups through culture. After living in settlements for thousands of years and evolving certain arrangements that allow us to do so, we can't tell our social and genetic adaptations apart."

To a far greater extent than is true of other creatures, we can modify our environment and behavior to reduce social stress. When the world is too much with us, we can shut the door, switch on the phone machine, or even get in the car and head for the hills. We are so various and flexible in the ways we adapt to different sizes of groups and settings that there are exceptions to just about every rule that social scientists can formulate. For example, the fact that prisoners in overpopulated jails do better in small single cells than in larger ones that must be shared suggests that under such conditions, too much social interaction is a more important consideration than lack of space. This theory would seem to be confirmed by the observation that college students in one study who lived in triple rooms did worse academically, were less altruistic, and had more psychological problems than those who lived in single or double quarters. A closer look at the data, however, reveals that the students so affected tended to be the odd men out in a three's-a-crowd scenario.

Perhaps the most significant difference between human and animal responses to crowding is that some of us, particularly thrill-seeking extroverts hooked on lots of stimulation, are tolerant of high social density to the point of enjoying it. Even those who find a teeming city hard going at first may adjust as they begin to enjoy the variety and novelty its vast population produces. Because the impact of high density varies so much with a person's natural or learned ability to handle it, Taylor says his studies of it have been "highly relative, showing both positive and negative effects."

New York's population of over 7 million may not explain the bad reputation it shares with other big cities, but even scientists loath to link high social density and various evils agree that coping with crowds is at least potentially stressful. On the simplest level, it is psychologically as well as physically restrictive, because being marooned in a sea of people undermines our sense of control—arguably the worst behavioral influence an environment can exert; simply telling subjects in one experiment that they could leave a packed room at will reduced their stress levels. Fifteen minutes spent in a gridlock or a checkout line is enough experience to enable us to understand how chronic exposure to such situations, which are far likelier in cities, can, particularly when combined with other environmental stresses, summon up debilitating feelings of learned helplessness.

Although numbers can lie when it comes to drawing conclusions about people's behavior in large groups, the presence of another person, or merely our belief in it, stimulates us as few things can. In an unusually inventive study, for example, psychologists observed that when someone else was close by, men in a small public rest room experienced physiological arousal, manifested by delayed, shortened urination. Partly because other people have this fundamental capacity to stir us up, our degree of arousal tends to climb with their numbers and proximity. This reservoir of excitation helps fuel urban phenomena from riots to a brisker walking speed: the bigger the community, the faster the average pedestrian's pace.

The Manhattanite's rapid clip has some advantages, but the heightened arousal that feeds it has a dark side. The overstimulation provoked by crowding can lower mood, increase withdrawal and aggression, and decrease helpfulness, particularly in men. These reactions would seem to be integers of the air of unsociability that is the major gripe the rest of the world nurses concerning city people. In a test of neighborly feeling among Pennsylvanians, researchers found that eye contact was very common in rural Parkesburg, less so in suburban Bryn Mawr, and rare in Philadelphia; another survey revealed that while fewer than 40 percent of urbanites shook hands with a stranger, almost 70 percent of small-towners did. Yet city people aren't meaner individuals per se than country folk. Studies indicate they are just as affectionate and accommodating with those they know; their cosmopolitan chill is simply a behavioral shield that defends them against the overstimulation of crowds of strangers.

Inside a big theater, if the designer has left ample room around our seats, we can feel festive rather than lost in a sea of other fans. The degree of comfort we experience in a crowd often depends on how well we are able to maintain our "personal space"—a kind of invisible bubble that surrounds the body. This unit of portable turf allows communication at close range while protecting us against the overly familiar or dangerous. The dimensions of personal space aren't fixed. In proximity to those we like best, the bubble shrinks; sometimes, the opposite dynamic obtains as well, inclining us to like a person, often a child, whom we've allowed to come closer than usual. If we're feeling angry, stressed, competitive, or formal, or find ourselves near the disfigured, smokers, or violent types, however, the bubble expands.

Judging by the considerable differences among cultures, the perimeters of personal space are also powerfully influenced by learning. Environmental psychologists have found that Americans of Northern European ancestry, along with the reserved English, generally extend their boundaries a cool eighteen inches from the body. But the Spanish and French, like the Greeks and Arabs, are comfortable at closer quarters. In a society made up of people from different cultures that have very different ideas about this intimate comfort zone, what one individual considers the right stance for a conversation may strike another as in-your-face provocation. Two parties needn't be from different ethnic groups to experience border troubles, either. Men tend to avoid close-up, face-to-face encounters unless they are ready for a fight; perhaps the side-by-side alignment they prefer accounts for the ubiquitous design of the bar. On the other hand, women are most comfortable head-on, and feel edgy when sidled up to. This different spatial orientation means that a woman who assumes a position that she feels promotes understanding can seem combative to a man, while the stance he takes to appear friendly can feel to her like a sneak attack on her flank.

Outside the bubble of personal space lie larger areas in which we take a proprietary interest, called territories. Anyone who has walked a male dog has witnessed an example of the way one species lays claim to certain resources and discourages rivals. The often stereotypical nature of animal territoriality suggests a biological etiology. Animals tend to preserve the same boundaries, for example, to maintain them for exclusive use, and to react aggressively to invasion. However, many backyards harbor an intriguing illustration of how learning influences even animal territoriality. East of the Rockies, purple martins no longer nest in the wild in their species' traditional way, but prefer to make their homes in gourds hung for them by appreciative humans. The birds' new territorial mode began as an adaptation to life alongside Native Americans, who liked having martins as neighbors because of their music, as well as their capacity for sounding alarms and eating insects. In turn, proximity to the Indians benefited the birds by discouraging predators and competitors for food.

Because a vastly more complex social evolution has powerfully influenced how people conduct territorial affairs, we are far more adaptable than other species. Rather than being rigidly restricted to a particular area, our turf is often far-flung, dispersed from home to office, car to club. If a newcomer ventures within our boundaries, we think twice before repelling him; should a stranger ring the doorbell, we may even permit him into the living room, while keeping more private spaces off-limits. Rather than relying on muscle, we usually depend on law and custom to help us hold our ground. A level gaze at a painting is enough to deter most fellow museum-goers from crossing in front of us, and simple markers, such as a mailbox or nameplate, mean that few will try to invade our primary territory, such as a house. Less personal "secondary" territories, say, a school or gym, can require more formidable markers, such as a security guard or Members Only sign, while public territories—a spot on the beach or a chair in the library—must be defended by occupancy or, for a brief period, a very particular kind of marker. In a bar, for example, a half-full glass will save a seat longer than a sweater draped over its back, because fellow patrons apparently assume that no one would abandon a drink in that setting. An even better bet for defending our piece of a public space in our absence is a friendly neighbor: 70 percent of moviegoers asked by a stranger to "save my seat" will do so vigilantly for up to fifteen minutes.

We also look to our territories to serve needs of a loftier order. Studies of hunter-gatherer societies show that a person's turf helps provide identity, privacy, intimacy, and protection from stress. One reason our homes are so precious to us—and being homeless is so debilitating—is that every time we cross the threshold, we wrap ourselves in a cozy, protective mantle of memories that helps sustain our persona. In the effort to extend the deep psychological meaning and comfort that our special places impart, we even invest particular objects with their values. The most obvious example is the flag, which conjures up the whole heartland in many breasts, just as a family snapshot or portable heirloom can turn a hotel room into a home away from home. Such territorial symbols not only help us unify our past and present, but also help us forge the future. In one study, researchers found the best indication of which students were likeliest to drop out of college was the decor of their dorm rooms during freshman year; the kids who embellished them lavishly and included local touches, such as university posters, were far more apt to stick it out than those who made little effort or decked their rooms with hometown memorabilia, say, a dried corsage from the high school prom.

If John Calhoun and his dark cautionary message of biological imperatives represent one end of the spectrum of opinion on our fitness for urbanized life, Ralph Taylor upholds the other—at least as an equal possibility. Concerning the old equation of big cities and behavioral sinks, he says, "If we're asking 'Does the urban structure make people do more lunatic things?' the answer is 'Maybe yes, maybe no.' A better question would be, 'Have we come to a point where the environment of the city is changing so rapidly that our process of cultural adaptation is really strained?' "

13

·

THE CRIME OF
THE CITIES

L IKE A river that meanders through a varied landscape, two long
blocks on Manhattan's Upper West Side that run from hard-
edged Amsterdam Avenue to sylvan Riverside Drive move past different worlds. A snapshot of the avenue's bustling corner might have
been framed in a Caribbean *barrio*. Big families spill out of crowded,
down-at-the-heels tenements onto noisy, littered sidewalks where
children play ball, women chat, and men play *salsa* or cards. A
photograph taken at the next intersection, just off the main artery
of Broadway, would show a prosperous scene of well-tended prewar
apartment buildings with marble lobbies and uniformed doormen.
While conditions on this half-funky, half-fancy block might be
termed "fair," the one directly to the west between Broadway and
Riverside is unequivocally "good"—a metropolitan wonderland of
fine townhouses festooned with Art Nouveau details. Beyond this
small pocket of livable blocks, however, those immediately to the
north and east are pocked by vacant lots turned into seedy parks,
unkempt public buildings, and little in the way of decent much less
attractive housing. These undeniably "bad" blocks are the perennial
settings for local misdeeds from car theft to murder.

When the uninitiated come here or to many similar polyglot neighborhoods in big American cities, they are taken aback by the precise suggestions about safe routes and places to avoid that seasoned residents offer. Although they are surprised that a territory whose boundaries can be walked in five or ten minutes can encompass urban heavens and hells along with purgatories, most soon learn to read the cityscape for signs of danger and oasis. Like urban veterans, they navigate toward a row of windows decorated with flowerboxes and away from boarded-up ones, toward a street of brightly lit row houses and away from the barren lawns of an apartment complex. Like scouts in the wild, experienced metropolitan travelers rely on their surroundings to help them figure out what is likely to occur where and to guide their own behavior.

According to Roger Barker's theory of behavior setting, we are apt to act in certain ways in certain places; the more clues a place provides about what we should do or not do, the more we will conform to them. Teenagers on their way to Riverside Park who stride across Amsterdam Avenue blasting their tape players usually turn them down once they cross Broadway and enter the bijou block. There, even small children intuit that a candy wrapper dropped on a sidewalk punctuated by blooming planters is anathema. The setting's ambience promotes observance of the proprieties where larger issues are concerned as well; the local cocaine wars, for example, take place on nearby bad blocks, despite a stronger police presence. Although twinkling gas lamps, marble steps, and tidy, flower-decked curbs alone can't defend against mischief makers and criminals, they radiate an almost palpable aura of the residents' concern with the social as well as physical standards of civility. On penetrating this invisible shield, the ill-willed immediately sense that here, wrongdoing will be met with individual and communal opposition as well as the official sort, and most go elsewhere to make trouble.

Two blocks away, on what is perhaps the most dangerous street in the neighborhood, things look very different. A big junior high school and its cyclone-fenced yard are an asphalt wasteland for most of the day. On the other side of the street from this anonymous expanse is a "vest-pocket park" whose broken fixtures and littered terrain repel all but vagrants and addicts. Beyond the park are several garages that, along with a few barred tenements, are the block's only privately owned buildings. This kind of no-man's-land setting, which supplies few of the physical cues that encourage good behavior, opens the door for the big trouble that regularly walks in.

Yet even this mean street and those to the north in Harlem were not always benighted. The process by which venerable neighborhoods, many of which were prosperous only twenty to forty years ago, have turned into slums involves myriad forces ranging from poverty to drugs to racism to disintegrating families. Among the afflictions that converged to wreak havoc in North Philadelphia, Bedford-Stuyvesant in Brooklyn, and Watts, however, the disruption of well-established behavior settings is rarely mentioned. Yet the boarded-up buildings and vacant lots of the inner cities are not just symbols of the disastrous change in the northeastern urban ecology but also active agents in its downward spiral.

Abandoned housing and other urban eyesores are the figurative and sometimes literal detritus of the legion of manufacturers and other employers who have abandoned the older cities for more profitable places to do business, from the Sunbelt to Taiwan. "When Bill Cosby grew up here in Philadelphia, his dad worked in a factory and probably made the equivalent of fifteen bucks an hour," says Ralph Taylor. "Today that guy can only get four or five bucks an hour at McDonald's, and the implications of that are enormous. For the first time in history, most of America's poverty is urban, yet the federal money for fairly effective poverty programs meant to address this shift, such as Head Start, has dried up. The cities are left to educate and otherwise service a huge, needy population that has few opportunities for advancement."

The socioeconomic plague Taylor describes has drastically afflicted the body as well as the soul of the city, physically pocking it in a way that, in a particularly vicious circle, aggravates the social sickness. "The places in which poorer groups grow up today are radically different than they were twenty-five years ago," he says. "A house's transition from a prime homeowner's property through the rental stage to abandonment happens much faster now. The huge volume of vacant housing that results has created an enormous opportunity for illegal drug activities, which is what most of it is used for."

Because vacant housing affords them plenty of room, marginal characters move in and take over a block's behavior setting from the established citizens that Taylor calls the "regulars." Once that begins to happen, the regulars can be pushed toward one of two unfortunate paths. Those who can afford to do so may panic and leave, giving way to unprepared, less committed newcomers and transients, which helps ensure the collapse of the setting. On the other hand, those who remain in the troubled setting by necessity or choice often try to protect themselves by retrenchment, thereby shrinking the amount of communal territory under surveillance. Instead of viewing the whole block as their turf, they may limit their involvement to their end of it, or even to their building.

The role played by abandoned housing in the decay of urban life is hard to overestimate. "Here in Philadelphia, at the urging of some ministers, former Mayor Goode sealed up eight hundred crack houses at great expense," says Taylor, "but there are so many empty buildings that when one drug place was bricked up, a new one could open right next door. If you combine a lack of jobs, enormous opportunities for 'social deviants' to flourish, fewer settings for 'regular' behavior, reduced public services, physical dilapidation . . . If I could do one thing to improve urban life, I'd rehab all the vacant housing."

As the ravages caused by crack houses and shooting galleries have shown, a community's troubles are apt to occur in the gaps in a behavior setting—spots in which the neighborhood's usual standards and degree of vigilance do not apply. That is why shady activities flourish not just in abandoned housing but on corners, in parks, or near bars, stores and other places where residents' control is limited. When these gaps expand, say, jumping from a candy store used for gambling to a nearby abandoned house used for prostitution and drug dealing to a vacant lot filled with trash, a deadly domino effect sets in. The worse the social climate, the more dilapidated the setting becomes, and vice versa. Before long, there are fewer symbols of the individual and group territoriality, from brightly lit doorways to clean sidewalks, that could help turn things around. Tragically, it doesn't take much—some peeling paint, an abandoned car, a boarded-up window—to set off the deadly chain reaction of fright, flight, or withdrawal on the part of solid citizens that destroys a once sound urban setting.

The more diverse the residents of a setting, the harder it is to band together to eliminate the gaps that invite bad behavior. Because a sense of community rests on shared values, feelings of belonging, and the ability to influence events, it is harder to achieve within an ethnically and economically mixed group. The members of a certain we-happy-few beach colony in Massachusetts are in such accord about the way things should be that all houses are painted gray and white, most of the dogs are descendants of a single black Lab bitch, and no laundry is ever hung outdoors. For similar reasons of homogeneity, things go smoothly in Hong Kong and Tokyo despite their huge populations. In melting pots like New York, Chicago, or Los Angeles, however, sharing a setting is a challenge, particularly since the acceleration in cultural diversity that began in the 1960s. For a pedestrian example, consider the parking problems experienced by one traditionally blue-collar Brooklyn block simultaneously undergoing both gentrification and an influx of low-income South American and Central American immigrants. At one end, the long-established families of white longshoremen would have nothing to do with their new neighbors, both groups of whom they saw as threats; they vigilantly defended the area in front of their houses by stabbing the tires of any "outsiders" who dared to park there. At the other end, the new gentrifiers were very much disturbed by the cars their foreign neighbors often parked in their driveways; to them, these vehicles were physical symbols of looming disorder and lawlessness—the thin end of the wedge. If something as seemingly simple and objective as parking can cause trouble among a diverse group sharing an urban setting, far more difficult problems can seem insurmountable, further encouraging aggression or withdrawal. "If you live on a block with lots of people you understand and get along with, things go fine," says Taylor. "But if your neighbors are really different from you, it's going to drive you crazy."

If a row of dilapidated and vacant houses creates a setting that begs for bad behavior, it seems reasonable to assume that renovating those buildings to their former Victorian splendor must be a solid step toward urban renewal. Despite the fact that the restoration of old urban neighborhoods is one of the few visible signs of hope in many cities, Taylor cautions that plugging in a few showplaces doesn't necessarily benefit a neighborhood. Gentrification means diversity, and that means not just clashes but anxiety, because we feel safest when we're near people like ourselves. In ethnically mixed neighborhoods, this fear of each other not only adds to stress but also decreases our sense of territoriality, which in turn increases the likelihood of trouble. "By definition, gentrification is a spotty process in which two very different groups—lower-income renters and higher-income owners—are thrust together. This polarity boosts the level of concern among those at the upscale end and dramatically intensifies the others' feelings of inequity. As soon as the rich people move in, you hear the long-term residents saying, 'Hey, I thought my kid would be able to grow up and buy a house here, and that ain't going to happen now. My taxes are going up. They're on my case about my cousin working on his car out front with the radio on. And where's the good old white bread in the supermarket?' "

In modern urbanized society, the greatest potential for misunderstanding concerns not race but money, says Taylor. "Can middle-class black and white people get on okay as neighbors? We don't yet know, although they seem to in many places. But there are chasms when poor and middle-class people, black or white, try to share a community."

The most chilling example of how the destruction of a supportive environment can accelerate a community's behavioral collapse may be the plight of the Central African tribe called the Ik. After they were displaced from their hunting grounds in the 1930s and left without an adequate material or spiritual substitute, the Ik culture disintegrated until, when anthropologist Colin Turnbull studied them in the 1970s, moral and social values were such that parents laughed when their own children were hurt. While the Inuit have traditionally coped well with extreme scarcity in the cradle of their homeland, the rootless Ik were unable to sustain the culture that sustained their humanity. Although the predicaments of two very different cultures can't be strictly compared, the negative influence of the setting plays an important role in the misery of the ghetto, just as it did in the decline of the Ik.

Seemingly planned to thwart a sense of belonging, the forty eleven-story buildings that comprised the infamous Pruitt-Igoe housing project in St. Louis finally had to be blown up in 1972. Its design was initially celebrated for the way it conserved space and reduced opportunities for vandalism. Because it offered none of the small, semiprivate communal areas that repel interlopers and encourage residents' territorial feelings, use, and surveillance, however, the development thwarted its tenants' psychological needs and the informal social networks that typically shore up the standards of urban neighborhoods. Sixteen years after its erection, the project was a nadir of depravity that begged for comparison to Calhoun's behavioral sink. By the time the towers were scheduled to be dynamited, rape, assault, and the damage to property—elevators were routinely used as toilets—were so widespread that more than half the buildings were vacant, especially their upper stories. At the very least, the nightmare of Pruitt-Igoe serves as a warning that when people's worlds are turned into wastelands, souls are starved and neglected as well as bodies.

As the Pruitt-Igoe debacle made clear, new-and-shiny isn't necessarily better, and good fences don't necessarily make good neighbors. The high-rise estrangement it epitomized can shadow affluent apartment complexes as well. A national survey conducted for the Federal National Mortgage Association in 1992 showed that 80 percent of Americans—particularly lower-income people and blacks and Hispanic Americans—still think that the single-family detached home with a yard is "the ideal place to live" and are willing to make substantial sacrifices to have one. Because of its high cost, however, home ownership has declined from 66 percent of the population in 1980 to 64 percent in 1990, which means that an increasing number of people are unable to secure something that is very important to them.

As "rugged individualists" distrustful of communal living, Americans have traditionally refused to regard an apartment as a "real home," implicating it in everything from the transmission of disease to impaired development. Even the first rich people to brave life in a Fifth Avenue triplex had to be wooed by the "maisonette," or town house built into a high-rise structure. Along with this antiapartment bias, residents of big tower developments, often set back off the street in nondescript grounds that discourage use, have to struggle harder to bond with hundreds of anonymous neighbors, keep watch, and even supervise their children. Not surprisingly, occupants of low-rise apartment buildings based on the personalized "real-home" model, grouped around smaller, semipublic yards, feel more territorial and neighborly than those who live in towers.

Although planners can identify a number of the circumstances that make for a vital neighborhood, rich or poor, the serendipity involved can elude programming, and the disadvantaged have not been the only guinea pigs for failed communal experiments. In the 1960s and 1970s, large tracts of rural land were turned into utopian "new towns" whose 15,000 to 100,000 middle-class residents were eager to sidestep urban hassles and suburban isolation. To reduce traffic congestion and pollution while affording convenience, the designers of Columbia, Maryland, planned things so that residents could shop within a five-minute walk or bike ride from their homes. It turned out, however, that many preferred to drive to larger, less expensive stores, which meant plenty of cars on the road and the bankruptcy of the smaller shops. In a similar burst of well-meaning but doomed hyperdesign, the planners decided to promote socializing by clumping an entire block's mailboxes in one spot. The citizenry complained, protesting that they couldn't run out to get the mail in their bathrobes. Although new towns do provide better facilities, people weren't any happier there, and by the early 1980s, many of the sixteen federally funded projects had been sold to private developers.

A major factor in the fallibility of utopian schemes is the gap between what we say we want from our surroundings and what actually fosters our health and contentment. Although the new towns were built on the assumption that tasteful buildings, attractive landscaping, and fine facilities lead to the good life, the reality seems to have as much to do with the familiar turf and rich network of slowly fostered relationships typical of a more venerable behavior setting. Although it might surprise the uninitiated, ghetto residents often have fond feelings about their neighborhoods. After razing miles of dilapidated cityscape, bureaucrats have discovered that appearances can be deceiving, and some very good things can be destroyed in the name of urban renewal—particularly cheap housing and complex webs of social support. Within what outsiders perceive as an unrelieved slum, convenient shops and services and the frequent crossing of residents' paths help shrink an impersonal city down to size, lower stress, promote socializing, and provide a forum for attacking problems too big for individuals to handle. Whether they're rich or poor, some neighbors can usually be counted on to watch out for the kids and the elderly, give practical advice, pull a group together, and last

THE CRIME OF THE CITIES ■ 197

but not least, smile on their daily rounds. In an unusually clever study that compared the positive as well as negative feelings of poor and wealthy city dwellers and suburbanites, researchers discovered that although slum residents had more complaints about their surroundings than the other two groups, they made the same number of benign observations. Their reaction helps explain why people stuck in "bad" housing may resist being relocated to impersonal new developments, and why, if forced to go, they often grieve: no fewer than 50 percent of a group of displaced Bostonians in one survey were still depressed by the move a year later, and 25 percent a year after that.

If outsiders find it hard to believe that residents of what seems like a ghetto might want to stay there, they find it even harder to understand why waves of newcomers struggle to join them. " 'Slum' is really a psychological phenomenon—a state of hopelessness—and despite what they look like, places labeled that way can be filled with hope," says Alaskan anthropologist Kerry Feldman. "In the Philippines, I studied 'squatter settlements' to see what influenced the poor to leave the inner-city slums for those farther removed—their 'suburbs.' After interviewing three thousand people, I found that those with the higher incomes moved. Now, we're talking about families who make fifty dollars a month, but that's different from thirty dollars. The little country villages these people had migrated from may look more picturesque to Americans, but to them, a squatter site isn't a slum, it's a step up. You ask, 'Why are you here?' and they say, 'For a job and a better education and life for my children.' That's pretty much what urbanization is about worldwide. If they can make enough money to climb to the next rung on the ladder, people are willing to live in a place filled with disease and crime."

If urban squatter and wealthy landlord, public-housing tenant and brownstone renovator are united by any response to the city, it is the dread of crime. Back in the fifteenth century, the Muslim Ibn Khaldun wrote that in contrast to the high-minded types bred in the open desert, the congested capitals of the West whelped depraved criminals. His bias persists, although, as Taylor points out, "People do plenty of bad things in the country, but Channel 10 news isn't there to record them." Of all the reasons people give for disliking big cities, crime tops the list, not without reason: compared with rural areas, urban rates of violence and murder are respectively eight and three times higher. The reasons for the discrepancy range from more poverty and bad role models, to more victims and goods to prey upon, to the decreased likelihood that the criminal will be caught and punished. But the physical setting plays an important, often overlooked role in the incidence of crime as well.

Crime rates are lower in places where the setting's regulars share and display a strong proprietary sense of territoriality. In certain urban, socially homogeneous working- and middle-class neighborhoods, for example, there may be little of the historical-society splendor found in more affluent districts that attract tourists, yet the atmosphere of communal involvement and concern is similar. Instead of bronze door knockers and marble facades, these humbler communities use neat houses, clean, well-populated streets, and planted borders to repel troublemakers. Even a birdbath or a statue of St. Francis helps send a strong message to potential miscreants that, as Taylor puts it, "if they try to climb in a window, they'll meet somebody charging through from the other side. If only one person on the block hangs a window box, nothing much will happen. But if many people do, and they also take action against trouble-makers, then you're talking. It's not the fixtures—the lights, the shrines, or the flowers—that prevent crime, but the social dynamics that drive and are driven by these environmental features. The phenomenon works in the opposite direction, too. As crime increases, the territorial markers that discourage it decrease, making things worse."

In posh suburbs, where just about everything is privately owned, markers abound and crime rates are lower. Poor communities, on the other hand, are often dominated by publicly owned developments and big stretches of no-man's-land where surveillance is minimal. Because most criminals come from such neighborhoods and usually venture less than two miles from home to do their dirty work, they particularly prey on these poorly marked and unguarded parts of their own backyards. Yet even in high-crime neighborhoods, pockets that encourage and display more territorial concern contend with less trouble: one housing project deliberately designed to encourage residents' supervision of common areas suffered only half the crime visited on another just across the street.

Crime is bad, but the majority of Americans will not be its victims. For them, the fear it inspires is worse. "We assume that our anxieties about being mugged or robbed vary in proportion to the number of crimes committed in the places where we live and work, but in fact, there's much more fear than crime," says Taylor. "This modern worry is far more widespread than the crime rate, particularly among the poor, black, female, and aged. It limits our behavior and generates stress. Among people who already feel beset, the specter of crime causes higher levels of anxiety and depression. This fear is not only a social problem, but also a mental-health problem."

Our settings play such a crucial role in our anxiety about crime that its level can usually be predicted by the physical deterioration of a neighborhood—really the residents' *perceptions* of deterioration, because different groups read those signs very differently. "The Main Line looks great to newcomers," says Taylor, "but people who have lived there all their lives see the changes wrought by urbanization, such as more apartments, fewer private homes, more people of various sorts, and less social cohesion, as being profound. From the opposite perspective, inner-city people, especially the young, might interpret graffiti, litter, and loitering as legitimate social responses to unfair treatment, just as a riot can be viewed as a political statement rather than a rampage. But to middle- and upper-class people, the mere trappings of incivility—spray paint that isn't removed or a window that isn't fixed the next day—spell risk. To them, these physical cues symbolize moral collapse, poor police service, and the regulars' lack of control over local behavior. A high-rent area such as Manhattan's Upper East Side can have plenty of robberies, but if there aren't any physical signs of deterioration, people don't move away, as they might in a less attractive area with the same crime rate."

Despite the deterioration of the urban setting, Taylor and others who study our interactions with it insist the future isn't hopeless. More than thirty years of their research have shown that modern urban environments can reinforce good behavior as well as bad; even in the much disparaged and potentially troubled milieu of public housing, for example, simple measures such as landscaping, the use of color, and the orientation of buildings to the street can help

increase the community's control and decrease crime. To environmentally minded behavioral scientists, part of the tragedy of the
cities isn't that no one knows what to do about them, but that we
know some things that could help yet haven't done them.

As more and more Americans flee the troubled setting of the cities
for the suburbs, they are pushing back the perimeters of the countryside. At the end of a long discussion of the influences of urbanization on behavior, Taylor adds another concern. "Although it's not
as visible as overt urban decay, since the 1980s the conversion of
rural land into suburbs has accelerated to a degree that affects not
just our well-being but that of all species. Even if only ten or fifteen
acres out of a hundred are perturbed, that can mean a radical disruption of a habitat, because breaking into a core area of some species
can make a whole territory dysfunctional. In South and Central
America, a species that only existed in a particular valley can be
wiped out. Nature is a human need, and we're wiping it out."

14

NATURE: WHAT HAS IT DONE FOR YOU LATELY?

L ITTLE FARTHER than a two-hour drive from Times Square flows a stretch of the Delaware River that Congress has designated as a National Scenic and Wild River. Long before it officially became a part of the American heritage, people had come to this country from all over New Jersey, New York, and Pennsylvania to go white-water rafting, canoeing, and fishing for shad and trout, as well as to admire vistas that rival those of the celebrated river valleys of France and Germany. Those with time to linger on the banks are occasionally rewarded with a glimpse of a bear or even a big rattlesnake swimming over from New York to Pennsylvania or vice versa.

Some residents, notably part-time "city people" and the same year-round families who try to reform the planning boards and schools, share the tourists' delight in the Upper Delaware region's rural quality and support the National Park Service's restrictions on its future development. But the assets and even livelihoods of many other members of the community—the lumbermen, farmers, and blue-collar landowners described by the city people as "the locals"—are bound up with what they can and cannot do with their property,

and they are outraged by the federal rules and regulations. The law
that means protection of wildlife and land to some citizens means a
loss of control and income to others. Rather than diminishing with
time, the locals' rancor has grown. The highway running parallel to
the river is decked with signs urging the NPS to "Get Out." Last
year's Memorial Day parade in the tiny hamlet of Fremont Center—
its 108th—featured a number of floats that had pointed Live Free or
Die and Don't Tread on Me themes, along with the usual anti-gun-
control tableaux. As these displays passed, the citified contingent,
who crave rural harmony as an antidote to urban friction, looked
away uneasily, afraid to boo yet unwilling to join in the cheers.
Neither they nor their ideological opponents are bad guys, and both
groups have legitimate concerns. The real problem is that they have
very different ideas about what nature is for. If it is to survive in all
its complexity, an awful lot of people from very diverse groups must
agree that nature is a mother lode of inner as well as material resources
that in some way enriches everyone, from the Sierra Club elite to
city slickers who never set foot in a park. Although we often overlook
or disparage as romantic the effects of natural stimuli on our well-
being, an expanding body of eclectic research shows that almost all
of us rely on nature—whether it is sprouting from a flowerpot or
stretching as far as the eye can see—to excite our senses, restore our
nerves, invite us to play, enhance our social bonds, and supply meaning and metaphor to our lives.

As the residents, tourists, and legislators of the Upper Delaware
region have learned, our species' response to nature is not the benign
constant some conservationists fondly imagine it to be, but a variable
that can fluctuate along social, ethnic, and economic lines. Even in
the last stand of the wilderness, only a few are able to enjoy its
pristine reaches. Because highways are few in Alaska, the rich charter
planes, while most citizens have a distinctly different outdoor experience than the sort afforded by the remote Brooks Range. "On a
nice weekend here in Anchorage, thirty thousand automobiles head
south on the only road," says Kerry Feldman. "It's a bumper-to-
bumper conveyor belt for two hundred and twenty-five miles until
you hit the end. Fisherpeople line up every ten feet along the streams,
some with TV sets. If you have some more money, you can go to
Mount Sisitna and hike around, fish, and pay two hundred dollars a

night to stay in a lodge, take a sauna, and drink wine with Europeans."

Those interested in probing for the underpinnings of our responses to the natural world before it became a money machine or an artifact of affluence often look toward the living testimony of another segment of Alaskan society. Of all Native Americans, the Alaskan groups have best maintained an unromanticized but profound connection to the natural world, one that continues to supply meaning to life as well as food for the table. In his work as a Bay Area psychologist and activist in protecting "sacred places" here and abroad, James Swan often turns to such tribal sources for remedies for personal and environmental problems. "My prescription for inner turmoil is spending a minimum of four hours alone in a natural area, with no activities or distractions," he says. "Just sitting quietly in that atmosphere allows most people to process a lot of emotions and issues they haven't been dealing with. Modern society focuses almost exclusively on the value of intellect and professionalism, which requires a lot of repression and denial. Both as individuals and as a culture, we can forget about some really important things, including identity and self-worth, so we need periods that allow us to get in touch with who we are and what really matters. We can also enhance the continuity between the conscious and unconscious mind by keeping track of the natural symbols that appear in our dreams, and reinforcing them by adding plants, animals, or whatever to our settings. We spend all but an hour a day indoors, estranged from the vast mine of meaning, art, metaphor, and teaching that we evolved in. In terms of the quality of life, that has got to matter."

The extent to which they continue to work that vast mine distinguishes traditional Native Alaskan societies from tribal cultures elsewhere. Before moving to Fairbanks, psychologist and University of Alaska dean Jerry Mohatt and his wife, Robby, spent fifteen years working and living on the Sioux reservation at Rosebud, South Dakota. If there is one striking difference between the lives of the two groups of Native peoples, he says, it is the Alaskans' continuing involvement with nature through the fishing, hunting, and gathering required by their subsistence life-style. "At reservations like Rosebud or Pine Ridge, the people live in a little town on small lots pretty much separated from the true natural environment. They might ride a horse or drive a car out there, but they are no longer working out their lives in relationship to it. Up here, nature, subsistence, and spirituality are still tied together much more intimately than in the Lower Forty-eight, where hunting and gathering rarely if ever go on this way. The mythological structure that gives meaning and a real deep sense of belonging in the Alaskan villages, which are small rural communities of twenty-five to several hundred people, is tied to the fact that the more subsistence-type activities you do in a setting, the better integrated you are with that whole social and physical environment."

One reason that hunting, fishing, and gathering are not just hobbies in rural Alaska is that goods are scarce there and can cost 40 percent more than they do in Anchorage. But subsistence activities also provide products essential to well-being that money often can't buy. In Feldman's 1981 study of subsistence beluga-whale hunting in small villages near Eschscholtz Bay, the most striking image is that of a freezer belonging to a mostly self-supporting eighty-year-old. In May, this man still had stores of salmon, caribou, rabbit, ptarmigan, sheefish, duck, muskrat, smelt—and one five-pound can of *maktak*, or whale blubber. Considering all the other stores, an outsider wouldn't grasp that this last item, or whaling in general, was particularly important. The Inupiat people don't feel well if they go without *maktak*, however; it is too precious a commodity to be sold or traded often. Like gold, it even serves as an economic standard by which other goods, such as gasoline, are measured. Because its worth transcends nutritional value, *maktak* is the status thing to have in your freezer, and in a year when there are no whales, or not enough, the people are deprived of something important, no matter what else is on the menu. "An anthropologist up here once suggested that the state run a whaling lottery for sportsmen," says Feldman. "The winners could hunt and the Eskimos could survive on food stamps. This guy left Alaska, but a lot of Western people think that way—'Give them food stamps.' To us, subsistence means supplementing what your wages buy with a few fish, but to Eskimos, a paying job supplements hunting and fishing. Subsistence is the system that gives them not just food, but self-esteem and spirituality. We find that hard to grasp because we try to isolate things—this is religion, this is economics, this is history. But really, they're not separate."

Just as infants are psychologically as well as physically regulated by their mothers, Native Alaskans are mothered by the country that provides the material, emotional, and cognitive context for their lives. Like contented babies or spouses, they are hooked on the deep, subtle satisfactions this environmental relationship provides. "The ways we become embedded not just in a human community, but in its physical environment, is something so deep that we haven't even studied it very well," says Mohatt. "To Native people, the village isn't just neighbors. It's the spirits of their dead, of trees, animals, and the earth. If you change the way people relate to that kind of collectivity, you're inviting trauma that's really disruptive of something deep. A person doesn't quite know why he's depressed or what he's lost, but he knows there's something wrong, something missing. It's very hard for us to grasp that."

As the behavioral consequences of our modern indoor life hidden from the sun suggest, technology poses insidious threats to delicate interactions with the natural world that our species has developed over centuries. Native Alaskan history shows that what might seem like a minor, even beneficial, change in a finely honed way of doing things has had disastrous repercussions, both psychologically and materially. "The conservationist ethic of Haida and Tlingit Indians disappeared at the end of the nineteenth century, when Western technology arrived in the form of football-field-sized nets to harvest salmon," says Feldman. "Instead of taking their fish traps out at night, as they had always done, the Indians started leaving them in so they could compete with the newcomers. This practice wiped out the fish in many streams and changed their culture. The same kind of process continues today. If you get, say, a snowmobile, suddenly you think about nature differently. The Eskimo people are hunting more caribou and shooting a lot more—the young people often just for fun—simply because they have this vehicle, and there are many less obvious effects. For example, some recent data from the Canadian Yukon Territory showed that Eskimo men are getting shorter. It turns out that the long hours they spend hunting and traveling over rough terrain in snowmobiles fuse their vertebrae. The men studied had also lost thirty percent of their muscle strength, because these modern labor-saving devices reduce the amount of exercise they get. The point is that you have to be careful what you invent, because it also invents *you*."

Proving that changes in the way people relate to nature alter the individual and collective psyche is trickier but no less important than measuring changes in bone and muscle. The suicide rate of young Native Alaskan men is very high—more than twice that of their white peers. On the other hand, Native Alaskan men over fifty-five virtually never commit suicide, while white men that age have the highest rate in the nation. Some public health officials tie the self-destructiveness of young Native men to the state's new oil-dominated economy, which has undermined the traditional male subsistence roles that sustained their elders; this challenge to the traditional Native way of life has been aggravated by the increased availability of alcohol, drugs, and guns, all of which boost the risk of impulsive violence. "Older Eskimos say that liquor wasn't really too much of a problem in the villages until fairly recently," says Fairbanks psychiatrist Irvin Rothrock. "Much of what they used to plan for and do for themselves is easier to come by now, one way or another. They don't need to catch as many fish to dry for their dogs, for example, because they've got snow machines." In the hard-working atmosphere of the summer fish camps, however, when the community still labors through the long glowing days to prepare stores for winter, no drinking is done. In recognition of the camps' atmosphere of sober traditional values, some twelve-step programs are deliberately begun there.

Many Native Alaskans continue to see their traditional way of life, worked out in harmony with nature and each other, as worth the forgoing of the glittering prizes of urban life. "Some Native people move to the cities, but the pull back to the village is very strong," says Rothrock. "An Eskimo might move south for ten years and hold a good job, then decide he needs to get back home to Barrow. That's kind of hard for me to understand, considering what it's like there in winter, but the village ties are powerful. Grandparents, brothers, and sisters will take over the kids if parents fail for one reason or another. Even the mentally ill are tolerated better than they'd be in a complex urban environment." Mohatt points out a particularly impressive manifestation of Native Alaskans' sense of rootedness. "The rural villages have many of the same problems that the inner cities do. You don't hear about too many people who grew up in the ghetto going back, but Native college kids here want to return to the village and help out. They feel a lot of pressure not to be uppity, and they worry about how to blend their education with village life. And like Native peoples elsewhere, they're very resentful of things written about their homes that make it seem as if nobody would want to live there. They love the sense of family and being tied to a community. When you leave a Native American community, you really miss it—we still miss Rosebud. A tremendous sense of kinship can co-exist with the same poverty, welfare, and alcoholism that exist in the inner cities, so that people here are not isolated in the same way urban people are. There's a supportive network of others who really care. They might not have any money to give you, but they'll help you one way or another if you need it—a place to stay, some food, maybe just somebody to sit with you."

Like countless cultures before them, Native Alaskans' ancient tradition of finding meaning in nature and living accordingly is now tested every day by a new economic reality: their own oil wealth. "Native corporations are the largest private landowners in the state, the people often have money, and their communities have the same tensions between conservation groups and those that favor economic development that exist between the Native community and the white society," says Feldman. "The question isn't how to preserve their culture, as if it belonged in a museum, but of how their culture will change, as all do, and of who is going to control its growth. It may be that Alaskan Natives will end up working all week so they can hunt and fish on weekends, too."

Those of us who live in urbanized society have long been estranged from our roots as hunters and gatherers and the bonds forged by sharing a natural environment. Yet we depend on nature, albeit in different forms, to provide inner resources as surely as the Inuit and Inupiat do. With his wife and fellow biologist Winnie Hallwachs, Daniel Janzen, a professor at the University of Pennsylvania who is the world's foremost tropical biologist, has spent the last six years helping Costa Ricans restore a tropical forest in remote Guanacaste province; his credentials for this job include almost thirty years' worth of research on esoteric dry forest biology, which has brought him two MacArthur Foundation "genius" fellowships and the Craoord Prize, biology's version of the Nobel. "Here's what nature does for us, no matter who we are or where we live," he says. "Human animals carry around this big brain, this big device for processing input. Part of our ability to use that device depends on the complex stimuli that challenged it throughout our evolution. Nature—whatever is out there, from a single tree to a whole forest—provides a big wad of the possible information we can process. If you diminish nature, you diminish the diversity of those stimuli. When we don't get input from nature, we end up having not much sense of smell, hearing, vision. Television becomes our reality. We can survive on that, and do. But it's not nearly as complex. I'll always put my money on the person who has the broadest background—I don't care whether it's in art, music, finance, or nature. I'm not arguing against cities and all they have to offer, but life is bigger than that, more than that. When we diminish nature, we turn off a lot of things in our own heads. People should care about nature for a very practical reason— the more experience we have, the better off we are.

"Over the past ten or fifteen years, I've been bothered by the fact that Americans think they're getting nature through TV—all those shows that bring the elephants and tigers right into the living rooms. This Muzak nature destroys the reality of people's experience outdoors. When they're actually in nature, it's disappointing, because the big, spectacular stimuli aren't coming as fast as they do on television. In nature, you might have to wait six hours or six days to see that bird. People who haven't been immersed in that TV background are much more affected when they visit the tropical forest."

While it is true that we could live out our lives in the artificial world of a spacecraft and see our "nature" on a video, the call of the wild, or what conservation-agency bureaucrats refer to as nature's recreational, spiritual, social, and aesthetic "noncommodity benefits," has a powerful allure. Americans spend $100 billion per year on outdoor activities and make 600 million trips to federal lands, which is a 60 percent increase over their patterns of twenty years ago. Our motivations for this remarkable turning toward the natural world are incredibly various, from mountain climbing to cooking over an open fire, bungee-cord jumping to breaking in the kids' new sleeping bags. No matter what our purported goal, however, study after study shows that we value the psychological effects of our outdoor experiences more than the activities per se. To anglers, peace and quiet, a sense of achievement, and socializing with peers matter more than fish, while hunters prize escape, exercise, and companionship over game.

Most of the people trekking through the wilderness these days aren't the rural Daniel Boone types of song and fable but well-educated, well-off people from metropolitan areas. "As urbanization increases, it's clear that many people aren't behaving as well," says Stephen Kaplan, a professor of psychology at the University of Michigan; with his wife, Rachel, he has spent more than twenty years studying how we're affected by experiences in nature. "If you just look at the statistics of spouse, child, and drug abuse, it's plain that there are a lot of pressures out there today that people have trouble absorbing. Nature could play a terribly important, although as yet almost unrecognized, role in reducing some of their stress."

Saying that nature "restores" us sounds nice but wishy-washy. The Kaplans decided to analyze this very common response to an outdoor experience and find out exactly what happens when we head for the sticks, hit the beach, picnic in the park, or otherwise unwind in natural settings. They found that nature works its specific restorative magic by easing a condition psychologists call "mental fatigue." This form of inner weariness and inability to focus sets in after a few hours, or months, of hard work that demands concentrated attention; among its symptoms is making the kind of dumb mistakes often labeled "human error," as well as irritability and unsociability. Because so many of the tasks that people in high-tech societies perform are very specialized, often involving long hours spent on one narrow, sometimes boring, activity, we suffer much more from mental fatigue than our ancestors did. When we immerse our weary brains in soothing patterns of natural input, such as rippling water, sighing branches, or drifting clouds, fewer things compete for our attention and drain our energy, and we start to feel refreshed. Even in a big city, we can enjoy this most obvious benefit simply by stepping from a busy street into the nearest park.

After many years spent studying extreme environments, Peter Suedfeld has boiled down the restorative essence of natural settings to this: "Urban places subject you to many stimuli, often of great intensity, that change quickly and continuously, while in the country, most change is gradual and periodic. And probably most important, there aren't nearly as many other people." Peter Hackett offers a dramatic illustration of this principle. "When you come back down from Denali, there's a big change in stimulation level. Human activity seems too much. The cars gross you out, and the noise. You need some time to get back into the scale of things down here. It helps if you can just sit on the river and enjoy the greenery and the fish and birds. You have a feeling of having been reborn into life."

When people talk about a restorative environment, they often refer to being in "a whole different world," says Stephen Kaplan. "Almost all mention the aesthetic dimension of nature, which must be very important, because we go to such trouble and expense to secure it." Even the person whose sole experience with nature consists of lying on a beach and watching the waves will not be surprised that those who visit the wilderness list aesthetics as one of their main objectives. Nature is not only a source of immediate physical beauty, but also a treasure trove of symbols and values on which we all rely. Whether framed by tent flaps or hanging on a museum wall, its manifestations stand for life itself, as well as growth, change, continuity, purity, freedom, mystery, and the transcendent. Even children's play reflects nature's symbolic magic; although children are more physically active in parks with hard surfaces, for example, studies show they engage in more fantasy in those that have some trees and grass. The other components of that "different world" we seek when we need to recharge our batteries include being away from our usual setting, feeling part of a larger unity, enjoying "soft," or effortless, concentration, and being in a place that suits our purposes. Apart from the natural aesthetics, notes Kaplan, "these conditions could be met in a basement workshop as well as on a mountain."

Kaplan's criteria for a restorative environment resonate with Mihaly Csikszentmihalyi's requirements for a so-called optimum experience. For many of us, recreation is our only opportunity to transcend our workaday limits, enjoy a sense of control, and escape our obsessions with time, personal problems, and others' opinions; also, being absorbed in natural surroundings seems to increase our chances for "flow." "The surge in the popularity of challenging survival programs and high-risk sports in natural settings is testimony of the substantial psychological rewards they provide," he says. "Piaget felt the ideal places to foster human development were both responsive and predictable, which is part of the reason why experiences in nature can be so satisfying. Leisure is a serious business that ought to be taken more seriously, because it helps us gratify higher needs that are hard to meet in more restrictive circumstances."

For ten years, the Kaplans monitored the responses of people temporarily freed from those restrictive circumstances by the "Outdoor Challenge" program, which afforded teenagers and adults, most with little previous experience, a nine to fourteen days' stay in a wilderness area of 17,000 acres in rural Michigan. Despite its name, the program focused not on Outward Bound–style heroics, but on immersion in nature and learning skills such as reading a compass and cooking outdoors; in addition, each participant spent forty-eight hours completely alone in a vigil called "solo." Long before such programs, tribal peoples sent their young males into the wilderness for self-affirming ordeals known as "vision quests." Like modern psychologists, these societies understand that in a physical setting in which we can prove mastery without fear of others' negative feedback, we are perfectly poised to shake off the learned helplessness that hinders inner change. "When you're in an atmosphere that offers few distractions and allows you to experience your ability to do things you weren't sure you could," says Kaplan, "it's much easier to figure out what matters and what doesn't, and to make some changes in your life."

After analyzing their data, the Kaplans found that the most striking results were the participants' sense of self-discovery, enthusiasm for the experience, and desire to make nature a part of their future lives. "Silence is a funny thing. I don't hear it too often. Last night I experienced the most I ever have," wrote one camper. Another recorded: "I don't understand it. I just feel so alive I want to yell and scream and tell everybody."

These responses to the Outdoor Challenge program and others like it are amplified in Hackett's descriptions of some of his more overtly challenging experiences in extreme natural settings. "Knowing you've been stressed physically, mentally, and emotionally and that you've come through puts you in a very powerful place," he says. "Near the top of Everest on my first ascent, I really should have been killed. I was alone going for the summit [29,028 feet, the highest point on earth], and I fell. Just as I was about to drop eight thousand feet, I accidentally caught a leg behind a rock and stopped, hanging upside down. It was a difficult situation to get out of. I was emotionally and physically spent—hypoxic, dehydrated, exhausted—but some incredible power took over. I was able to think extremely clearly, and to do climbing that normally I wouldn't even think of. Under just the right stress conditions, you can reach this peak sensation of feeling one with the universe. I would not want to get that close to death again. That would be pushing my luck. But I often think about whether there are other ways I could harness that kind of power. Feeling at the peak of your powers as a human being can change you, especially if it happens repeatedly. There's a parallel with international travel—once you've experienced it, your perspective is different for the rest of your life."

When we expand our perspective by traveling to some remote wilds, most of us take someone along for the ride. Perhaps the most surprising revelation in the statistics on the use of wild places is that despite what people say, the vast majority go to the wilderness to be with others: fewer than 2 percent of the visitors to the big parks, for example, spend their time alone. As a result, summer visitors to Yellowstone are unprepared for traffic jams, supermarkets, wine lists, and "wild" bighorn sheep begging for nibbles at car windows, and still less for elbowing their way to natural wonders through hordes of their countrymen. After a day or two that can only be described as suburban, the more solitary souls realize that there are two Yellowstones, and the one they're after requires a bedroll. "For some of us, the range of emotional and physiological response nature elicits is greater than what other people provide," says Winnie Hallwachs, who spends half of each year in a remote Costa Rican national park. "This means our experiences with others and with nature tend not to overlap too much, becoming an either/or choice."

What most people want when they get fed up with the madding crowd and head out of town is "selective solitude." Rather than true isolation, they simply seek more control over which and how many others they have to deal with. While they enjoy being with their own party, for example, they dislike meeting strangers on the trail, which generates a peculiar sense of being crowded in some very uncrowded places. Because most visitors aren't solitary nature lovers, but sociable middle-class citizens seeking companionship, intimacy, and the sharing of experiences, planners increasingly think it is foolish to insist that a pristine setting must be a park's main attraction. Instead, they try to offer visitors a mixture of sites—some easily accessible and others remote—geared to handle the conflicting needs of backpackers and RV owners, families and singles, athletes and the sedentary. "Along with pristine areas, parks need beaches and educational programs," says Daniel Janzen. "They must be user-friendly to survive, because people won't support what they don't understand."

If an increasingly urbanized population is to understand and support it, our definition of "nature" must include not just vast stretches but small bites of it, from the trees on our sidewalks to backyard gardens. "Although the best way for people to maintain an even keel is by having a mixture of short- and long-term restorative experiences, it's a serious error to equate wilderness and nature," says Stephen Kaplan. "Relatively small pieces of it, even a single tree, can have great meaning for people."

We're surrounded by proof that we don't always have to go outside to experience some of nature's benign effects. Even the most dedicated *boulevardier* includes a pet or animal figurine, a plant, or a landscape in his apartment's decor. Yet by far the most popular scenes in most homes are the ones framed by windows, especially when they include a tree. "Views are tremendously important to us," says Kaplan. "They must be, because we pay more to have them, and make a lot of effort to manage them." People who look out on trees and flowers in the workplace feel less pressured, are more satisfied with their jobs, and suffer from fewer ailments such as headaches. A study by Marc Fried that focused on identifying the elements that determine our quality of life showed that although the strongest predictor of satisfaction was a good marriage, the immediate surroundings, especially the natural environment, came next. The lower the subject's socioeconomic status, the more that was true. "Sometimes," says Kaplan, "what's right there is all a person has."

Like a big mountain, a small garden stimulates, restores, and delights us, just as it poses challenges, promotes mastery, provides exercise, and relieves monotony. When tenants in New York City public housing developments were given plots to cultivate, their pride, self-esteem, and degree of sociability increased, and they even enjoyed some benefits that their solitary rural peers miss out on. Not only did their level of satisfaction with their whole neighborhood rise, but the new territorial markers their labor provided also reduced indoor and outdoor vandalism. "It's all too easy to assume that people who live in the inner city, where there's less nature in evidence, don't need it or care about it," says Kaplan. "That's particularly unfortunate because nature experiences could play a vital role in helping people manage the problems of that setting. The urban poor are under an all-out assault that makes it difficult to focus one's attention, which in turn makes all sorts of other things harder. They need restorative experiences. Is a garden nature in a microcosm? We really don't know, but an incredible number of people of all sorts garden. One thing that they all get out of it is experience with plants, water, trees—things that will always be. Things that are as close to being universals as we can find."

Fostering a global appreciation of the importance of those universals to our well-being is a vital step in securing their future. "It's terribly important to realize that within all people there's a very strong pull toward nature and also a fear of it," says Kaplan. "How those two things resolve themselves varies with time and place. Although the more experience a person has with nature, the stronger the pull toward it, I argue that our species has a general preference for and valuing of natural stimuli, because with a very few units of experience, a great many people find a natural environment to be profoundly affecting. When people who've had little previous outdoor experience say they want to make it an important part of their lives in the future, as often happens following participation in a nature program, that's compelling. In a short time, something important happens to all kinds of people in natural places."

15

.

SAVING THE
WORLD

I F THERE is one experience of nature that unites just about everybody on earth, it is what Gordon Orians, a professor of zoology at the University of Washington in Seattle, describes as "how crazy people are about flowers. We all respond to them, but why? When evolutionary biologists observe that a species has a strong preference for something, they ask, 'Why was that attraction advantageous?' My speculation is that flowers have always been a very important information source for us. In the species-rich tropical settings in which we evolved, it's hard to identify a plant whose fruit you might want to eat without a flower to flag your attention. Blossoming also provides very important cues that help us predict the timing of important events in nature. Then too, we human beings are very fond of sugar, and until recently, our only source was honey, which comes from flowers. Beekeeping and robbing nests for honey are very old practices. Even today, Africans are led to bees' nests by honey-guide birds, who then wait for the remains of the honeycombs. For all these reasons, paying attention to flowers probably had an enormous pay-off for our ancestors."

Just as nature affects our behavior, our behavior affects nature. Preserving not just this or that forest or wetland but nature in all its complexity requires a worldwide consensus that experiencing it is a human need transcending cultural boundaries. One way to secure that agreement is to demonstrate that diverse individuals and groups share widespread preferences for certain natural things, such as flowers. Although this benign premise strikes nature lovers as self-evident, it is the subject of heated academic debate. Some scientists, including Orians, argue that our attraction to trees and waterfalls, animals and blossoms is inscribed in our genes; others claim that such preferences, when they exist, are primarily learned. "I'm not sure that one can say people do better in natural surroundings because we evolved in them," says Peter Suedfeld. "I'm not even sure we do better in natural surroundings. That's why we have cities. Along with many others, I like mountains and oceans, but we sure didn't evolve there, nor as solitary wanderers through the forest. We're in the midst of a neo-Romantic movement celebrating oneness with nature and wilderness and Gaia and all that, but a lot of the excitement is more verbal than anything else."

Even scientists convinced that our species has an innate preference for natural things agree that many individuals or groups don't show many signs of it. "People don't like to feel helpless or scared, and nature can certainly make us feel that way," says Stephen Kaplan, "especially if we're not equipped to deal with it." His remark particularly applies to the largest group resistant to nature's charms: children. To bolster their argument that nature loving is not inborn, some behavioral scientists point to the fact that children favor nature's tamest manifestations—if they favor any at all. Asked to choose among natural scenes, most pick pastoral views, avoiding the jungle and especially the desert; even benign trees and grass are the lagging ninth and tenth choices on their lists. When they draw, their own houses and streets are by far their favorite environmental subjects. Nor do children necessarily prefer to play in natural settings; one study conducted in a housing development showed that only 2 percent of all recreation occurred in the parklike areas provided. Equally concerned about isolation and unboundedness, children from rural Vermont are as anxious about venturing into the woods as those from the inner city. In fact, the ranges of urban, suburban, and country children extend about the same distance from home, and those of rural girls and younger children are much more limited.

Along with age and personal taste, nature loving can vary with ethnicity and class. According to Mihaly Csikszentmihalyi, it is one of the "finer things" cultivated only after the fundamentals have been taken care of. "When I came to Chicago from Rome about thirty years ago, it took me three or four years to feel I was living in a tolerable environment," he says. "I interviewed some immigrants from the Greek islands who had settled in the slums, because I was sure that if anyone would be aware of a diminished quality of life in terms of the environment, they would. But it turned out that they were so intent on making money and moving up that they were oblivious to the fact that they lived in an awful place. Our current enthusiasm over the aesthetics of environment and the romanticization of nature is related to the much larger numbers of educated, affluent people who have the luxury of enjoying such things. People trying to survive in a garbage dump in Mexico City don't care about beauty and nature. For them, it's a luxury to imagine living in a better dump—perhaps one reserved for trash rather than garbage— let alone in a pristine setting. It's unfortunate, but basic needs must be met before people can worry about the environment."

According to Orians, the fact that our reactions to nature have a strong social component doesn't rule out a genetic preference. "To understand children's attitudes, you have to think in terms of development," he says. "At what point in life does it pay to start responding to a thing? A very small child wants to be near mother, period. For some time, he has little motivation to be interested in the larger environment." After a respite, youthful *ennui* with nature again peaks in the teens, when the social, not the physical, setting is of premier importance once more. Among primates, adolescent males must leave their troop and join a new band; since this "dispersal behavior" serves the important function of minimizing inbreeding, there is no advantage to an attachment to the home setting. That similar feelings stir in our own young can't be proved, but Orians is not alone in observing that "traveling through a scenic landscape with teenagers can be frustrating. But eventually, there's that day in the car when one of them says, 'Wow, look at that scenery!' and you practically drive into the ditch. The landscape has suddenly become relevant." Then too, says Kaplan, "when we're young we have a shorter attention span and are very interested in action and excitement. As we get older, we become more interested in peace and reflection, and nature becomes more attractive to us. In the studies we've done of gardening, the most avid fans have often been adults who were forced to do it as children, when they absolutely hated it."

His researches have led Orians to conclude that scientists should test the hypothesis that *not* having a preference for nature is the learned behavior. "People from inner cities will score natural environments lower than people more familiar with those settings, and whether you were scared by a snake or not when you were a kid does make a difference. But our reactions to nature spring from a combination of genetic composition and what life has hit us with so far. The real question we should be asking is whether experience is the only influence on our responses to nature. If not, what else is involved? This is not a radical idea. Your height has a strong genetic component to it, but it also depends on your diet. That's why Japanese people who grow up in America are taller than their relatives in Japan. The fact that we speak English has nothing to do with genetics, but the fact that we can talk does. Genes and environment are involved in our behavior and structure. The nature/nurture debate is a dead end. The important question is, 'What are their roles?' It would be extraordinary, considering all the genes and chromosomes we have that influence everything we do, if our attitudes and responses to nature weren't affected by them."

Assuming that our species does indeed have general preferences for and reactions to certain natural things, Orians believes that this behavior is best viewed in the context of what biologists call habitat selection. "This process of choosing the right places in which to do things is a universal activity among organisms and something our ancestors did all the time," he says. "When you encounter an unfamiliar environment, first you ask yourself if it's worth exploring. If so, you do some reconnaissance—you learn about it. Then comes the actual decision to use it. Over time, individuals of a species will come to prefer—even enjoy—the elements of natural environments that have increased their ability to function, and that pleasure motivates further awareness of those stimuli. People don't have sex to pass genes, for example, but because it's fun. Those who didn't think so were consistently underrepresented in the population. We find things such as sunsets and thunder and lightning so compelling because we once benefited by attending to these ephemeral cues. Basically, the sunset means 'Are you where you want to be when it gets dark?' The arousal that stimulus evokes can be either pleasure or panic. On the porch of the cabin, holding a martini, you find the sunset enjoyable. In remote open country, you had better pay attention to that red in the sky and get to your destination fast."

According to this evolutionary perspective on our reactions to nature, Homo sapiens is, like other species, inclined to favor the environment that helped to write its genetic and behavioral script. During our long evolution as hunters and gatherers on the East African savannah, we developed a taste for the kind of terrain we continue to prefer in paintings, drawings, and photographs. We like natural scenes best, even if their artistic quality is poor. A meandering creek, a thundering waterfall, or any other body of water makes a picture the hands-down first choice, which isn't surprising, considering water's importance to our survival. Next come the savannah-like vistas, such as an open meadow punctuated by trees and bushes, that would allow us both to gather information quickly and to seek shelter, and also piques the interest of an incorrigibly curious breed. Kaplan agrees: "In studies of preferences for natural things, the bottom line is that we're attracted to widely spaced trees and smooth ground textures that allow easy locomotion. If a setting is forested, we like open spaces in it, and if it's prairielike, we like some trees. This preference is consistent across cultures, but there are differences among groups. Inner-city Detroiters, for example, prefer a more orderly version that might include, say, a park bench or shelter." According to this "savannah hypothesis," we carry some of our East African heritage wherever we go. "People recreate open parkland all the time," says Orians. "Around the world, gardens reflect the same sensibility."

It has been said that man is the species who, where are no trees, plants them, and where are trees, cuts them down. Some of Orians' recent research probes their special place in our hearts, which has been demonstrated many times across cultural boundaries. City-dwellers prefer a leafy view to a lavish lawn, for example, and despite the potential of thousands of dollars of easy profit, most owners of small woodlots aren't interested in even a judicious thinning of their timber. "When I ask people from different cultures—North America, South America, and Australia so far—to look at photographs of differently shaped trees from an East African species, they all rank certain features highly," he says. "They like the tree to be wider than it is tall, and for the trunk to split low, close to the ground. I speculate that we find these elements attractive because they increase our potential to function. They provide better shelter, for example, and are more climbable." One ubiquitous example of our affinity for this broad, spreading, user-friendly tree is the *bonsai*, says Orians, and such examples of the way we manipulate nature to produce beauty provide him with important information about our inherent preferences.

"If you ask people what could be added to a landscape to make them like it more, they invariably want some water, which was extremely important on the savannah in dry season. Japanese rock gardens, incidentally, are really seascapes. They're raked to simulate water moving around the larger rocks, which are islands. The other feature people mention are big mammals, the ultimate attraction for hunters. When Humphrey Repton, the English landscape architect, showed his prospective clients 'before' and 'after' views of their properties, he usually put some deer and sheep in the 'afters,' along with a body of water. Nature does not just mean pristine wilderness, which isn't what most of us prefer or try to simulate when we make aesthetic natural environments or works of art."

If we have a preference for natural things in life as in art, we are likely to express it in our own backyards. Despite the recent upsurge of interest in the more macho outdoor pursuits, the National Gardening Association reports that no fewer than 80 percent of the country's 93.3 million households delight in growing green things. This tendency to favor pieces of nature that bear our thumbprint has been the rule throughout Western history. Only recently has a sizable group begun to advocate conserving the same wild resources our species has traditionally struggled to exploit. Far from being new, the conflicts that threaten environmental quality today reflect centuries of ambivalent and often hostile feelings about and behavior toward the natural world.

After the 1989 earthquake rattled the San Francisco Bay Area,

Stanford researchers conducted a study that helps us understand a type of experience with nature that must have been common among our ancient forebears. When the psychologists analyzed the responses of over a hundred people to the cataclysm, they found that during and shortly after the quake, their subjects experienced changes in their sense of reality and time, as well as in cognition, memory, and anxiety level. The team concluded that traumatic natural events can "profoundly alter the psychological experience of non-clinical populations and are frequently related to dissociative phenomena [the fragmentation of a person's experience into autonomous units]." Although our scientific understanding and technology shield us from the psychological and physical effects of many such disasters, we can hear the echoes of our ancestors in the fatalistic voices of modern farmers faced with a drought or of residents of flood- or twister-prone areas. Merging the natural with the supernatural, at least metaphorically, they might say with a shrug, "What can we do? We just hope for the best," or "It's in God's hands."

For most of Western history, people's mixed feelings about nature were augmented by elements of the Judeo-Christian tradition and later by the economic imperatives of the emerging bourgeoisie: for the glory of God and the family fortune, civilized people had the right, even the duty, to subdue the unruly wilds and the "heathen" who inhabited them. During the Middle Ages, vast stretches of the untamed European forest were regarded as living symbols of chaos and danger that teemed with demons and monsters. Just as evil lurked in raw nature, good was to be found in civilized communities and gardens. Something of this medieval apprehension lives on in fairy tales of monsters in the woods that ravish women and eat children, as well as in campfire yarns of werewolves, Bigfoot, and the Loch Ness Monster. Traveling from monastery to monastery, the twelfth-century French religious reformer Saint Bernard of Clairvaux ordered the shades of his carriage drawn so he could be spared the ugliness and disorder of Alpine landscapes that enchant tourists today.

Even within the much maligned European tradition, however, other more harmonious approaches to nature existed as well. Francis of Assisi not only talked to the birds and addressed the sun and moon as relatives, but befriended the dreaded wolf of Gubbio, whose story still delights children eight hundred years later. As Kaplan points out, "You can still see that respectful religious attitude toward nature in the contemporary Amish farm culture." Something of this mellower disposition was expressed by Thomas Jefferson, a classic gentleman naturalist whose diaries reflect a desire to understand and cooperate with rather than conquer nature, and who envisioned the ideal citizenry as educated farming families. It is no accident that Monticello, the prototypical upper-middle-class American dream world, evolved in the temperate, well-behaved environs of Virginia. Along with their harsh brand of religion, the Puritans who settled to the north in New England were burdened with cold winters and poor soil; not surprisingly, they and their westward-bound descendants viewed the natural world as a hostile place filled with indigenous savages. The middle Atlantic and southern settlers, often more easygoing Anglicans, Quakers, and Catholics, were blessed with gentler lands and climates. Many were also better educated and richer than the dour Puritans—two attributes epitomized by Jefferson still associated with an appreciation of nature. Although he charged Lewis and Clark with charting the wilderness, Jefferson and his contemporaries did not find such unmodified natural settings beautiful and worthwhile per se. Like the many Americans who live in suburban approximations of his bower-filled vision, Jefferson preferred pastoral vistas that combined woods and lawns, deer and cattle.

Not until the wave of literary and artistic Romanticism washed over Europe did wilderness become fashionable. As Csikszentmihalyi puts it, "Not even the Swiss cared much for the Alps until nineteenth-century English clergymen began writing about them." The parsons' sentiments were prompted by early signs from the Industrial Revolution that pristine nature was in increasingly short supply. To realize their fantasies, European Romantics had to turn to the vast frontiers of America. There, her own artistic elite, including James Fenimore Cooper, Thomas Cole, Albert Bierstadt, and Henry David Thoreau, celebrated nature as cathedral. The notion caught on with their pragmatic countrymen. Lacking a Chartres or St. Peter's, Americans deftly adopted Yosemite and Yellowstone as national shrines and sources of pride.

From the moment when the wilderness was first considered a source of aesthetic and spiritual as well as material wealth, there have been heated battles over what to do or not do with it. After the Civil War, Americans began to recognize that even their vast natural resources were finite, and to consider how they might be managed and conserved. In 1864, Yosemite Valley in California was set aside as a national monument, followed in 1872 by Yellowstone, the world's first national park. To the legislators who preserved them, one of the most appealing features of these beautiful places was their lack of conventional commercial value.

As the continuing battle on the last American frontier of Alaska proves, conservation would not always be so easy. Indeed, even conservationists battle with each other over the question of what nature is for. The lines of their classic feud were drawn at the turn of the century by naturalist John Muir and forestry official Gifford Pinchot. To Muir, nature was a spiritual and religious resource that restored the individual who subjected himself to it. To Pinchot, it was also a material resource that should provide the greatest good for the greatest number—ranchers, lumbermen, hunters, and, later, large families in RVs, as well as solitary backpackers, birdwatchers, and biologists. Much conservation today is cast in the Pinchot mold, aimed at timber production and watershed management as well as preservation of pristine conditions. But when lawmakers must choose between money and wilderness, economic concerns often prevail. As the decision to build the pipeline through the Alaskan frontier proves, where conservation is concerned, being almost a virgin is considered pure enough.

As the subdue-the-earth philosophy is losing some of its appeal in certain strata of the modern West, supplanted by the "green" movement, it is tempting to simplify and idealize the wilderness and the "noble savage" who lived there worshipping nature and killing only what he could eat—after making the appropriate apologies, of course. When we indulge in this form of romanticization, however, "we're likely to get a lot of things wrong," says Kaplan. "For example, at the time of the Mongolian invasion of the American continent, which started at the Bering Strait and moved straight down south, many species disappeared. There is speculation that when the invaders ran out of space and decimated much of the game, they had a huge population crash of their own. Seen in this context, the reverence for nature evident in the cultures of their Native American descendants is a practical behavior provoked by a very tough experience. There are arid areas in the Southwest that support very few people today but once accommodated thousands and thousands of Native

Americans, because they dealt with the setting and sparse rainfall so well. It was *adaptive* for these peoples to revere nature."

If it is to survive the onslaught of our very different society, we must, like the early Native Americans, see that it is adaptive to value nature, for both psychological and material reasons. In March 1989, eleven million gallons of oil spilled into Alaska's Prince William Sound, despoiling a remote area of the earth renowned for its wilderness and wildlife. Along with its ecological value, the region also yielded tens of millions of dollars in fishing revenue each year. When psychologists at the local Valdez Counseling Center conducted a yearlong three-phase study of the spill's effects on the psychological well-being of the nearby communities of Cordova and Valdez, they found that most of their 117 randomly selected subjects had suffered as a result, particularly from depression and the symptoms of posttraumatic shock. Their report concluded: "The oil spill was an extreme stressor that could cause emotional problems for most area residents. Cordova [the town that had more fishing] was found to have a higher incidence, intensity, and duration of stress as a result of the spill than was experienced in Valdez. Evidence of delayed and cyclical stress reactions was found, as well as a cause-and-effect relationship between stress and the incidence and severity of depression."

No one on the planet who had access to television was unaware that the Prince William Sound oil spill was an environmental disaster. Watching the images of greasy beaches and wild creatures struggling with deadly goo, many viewers felt twinges of the same hopelessness and helplessness that the area's residents experienced. Unlike the Alaskan oil spill, however, most chronic environmental problems are not particularly photogenic or immediately stirring. A report by the National Acid Precipitation Assessment Program, for example, shows that our "national vista"—the distance we are able to see—dimmed 25 percent between 1948 and 1983; because a haze caused by air pollutants can drift hundreds of miles from its source, that cloud obscuring the view of the Grand Canyon could be from Los Angeles. Even though air pollution spoils our views, accounts for $250 million in health costs per year, and affects about half of the people in the country, we continue to burn with abandon the fossil fuels responsible for most of it. One reason we do involves a human tendency that has generally been advantageous. "Because Homo sapiens is an adaptive animal, we can make do in almost any setting, physical or political," says T George Harris, the behavior-minded editor of the *Harvard Business Review*. "The danger is that we'll tolerate worse and worse air and water until things have gotten so bad we feel we can't do anything about them."

Trying to draw sensible battle lines for the war to save the natural environment is a daunting task, not least because of all the psychological processes our efforts involve, from perception to decision making. Paul Stern, a psychologist at the National Research Council of the National Academy of Sciences, in Washington, D.C., is the study director of its Committee on the Human Dimensions of Global Change. He spends a lot of time thinking about how behavioral issues, both individual and collective, affect environmental quality, and a conversation with him is not altogether upbeat. "The first problem with making a sound decision in terms of the environment is to figure out what's worth doing," he says. "It's often not easy to differentiate the important choices from the unimportant ones." Stern uses energy conservation in the home as an example. Surveys show that systematically, we tend to be wrong when we try to guess how much energy we use; we overestimate the power required by things we can see or hear or have to do over and over again, such as turning out lights, and underestimate what is gobbled up by things that work automatically and aren't easily perceived, like the water heater in the basement. "Even with some consumer education, this behavior is hard to change," says Stern. "I'm not against turning out the lights, but if you think that's a significant form of conservation, especially in comparison with having an energy-efficient water heater, you're fooling yourself."

Even if we have identified a bit of behavior that can really make a difference to environmental quality, it often takes more than good intentions to make us implement it. "The important things we can do to save energy tend to cost money and happen infrequently," says Stern. "We're not likely to replace a working water heater just to help the environment, unless we're also going to see big savings over a short period or a substantial rebate. The last time I bought one, the plumber brought a pretty efficient model, because the power

226 ■ THE POWER OF PLACE

company had a rebate program to get consumers to install large electric heaters that could be timed to operate at off-peak hours. It was in their interest to do this. Otherwise, they would need new power plants, which the Public Service Commission didn't want them to build. If the consumer went along with the plan, he got a refund of a hundred and forty-five dollars and an environmentally sound product. But it took a lot of parties to play out this nice story, which benefited the environment, the power company, and the consumer."

Of all the ways of helping people think in terms of "environment tomorrow" rather than "me today," policies aimed at the purse are the most effective. As Stern's water-heater example illustrates, the big hurdle in saving the environment isn't so much our thoughts and feelings as our deeds. Even consumers who don't know or care about environmental quality improve it when they buy an efficient water heater to cut their fuel bills and get the rebate. "There's many a slip between attitudes and behavior," Stern says. "I don't quite mean that it all boils down to economic incentives, which don't work if they're not implemented effectively. But cost is an important factor in doing the right thing—it makes you think twice. In 1980, we did a statewide survey of energy use in Massachusetts. People's beliefs mattered a lot when it came, say, to thermostat settings, which is something that saves them money and is easy to do. But philosophy didn't matter when it came to insulating walls, which saves more energy but is expensive and inconvenient to do."

No one could be surprised to hear that people are likelier to behave more responsibly when a material as well as moral reward is involved. Stern and his colleagues have already learned that individuals generally conserve energy in the home in proportion to what it costs them to do so and the money that they will save. Now they're in the process of gleaning this same kind of pragmatic information about behavior at the institutional and political level. "When people make an environmentally wrong choice, they usually do so for a good reason," he says. "If you care about the environment, you have to care about politics, because many of the reasons why so many people are doing so many things wrong have to do with policy-related barriers to doing the right thing, from the low price of fossil fuels to the encouragement of suburbanization by the Highway Trust Fund's road building. No one jeopardizes the environment because he wants to. People do it for a benefit they don't want to give up. To save the spotted owl, the government decided to limit logging in its habitat to more acres than the lumbermen wanted and fewer than the environmentalists wanted. How you resolve that very common kind of situation is a political question. The people who cut down the trees may love the natural environment, too, yet they do things the city people say are shortsighted because they need to make a living, and that's the only work they can get. If they want that behavior changed, the city people are going to have to help solve the country people's economic problem, whether it's here or in the rain forests."

Although talking to Stern about human behavior and the natural environment has its dark moments, he admits that there have been some encouraging developments. "You remember the oil fires in the Cuyahoga River? We don't have those anymore. The reason we've had some success in improving our water is that there were relatively few significant sources of river pollution, and government agencies could force those industries to change. When I was growing up in New York in the 1950s, we used to watch the raw sewage float into the river. You flushed your toilet, and it went right into the water. Today that doesn't happen, and there are lots of fish in the Hudson again."

Just as the Mongols who invaded North America and their descendants may have learned by bitter experience to conserve and respect nature, modern societies are increasingly motivated to do the same by the psychological as well as economic price tag of technological progress. "There's more and more evidence from many different areas that we resonate with natural things," says Orians. "Surgery patients who can see trees from their hospital window recover faster and with fewer complaints than those who look at a brick wall. People who work in places that have no windows go to all sorts of extremes to try to simulate them, and the amount of material on their walls and its content is different from that in offices that have windows. Within clinical psychology, there's a whole new field of therapeutic horticulture—some prison reformers are even trying to include animals as well as gardens in that environment. Although it jeopardizes nature, suburbia itself is an attempt to recapture it, and a lot of people go to a lot of trouble to live in its greenery. If the hypothesis that we have a genetic preference for natural things is right, increasing urbanization means that we'll experience all kinds of behavioral consequences and deficiencies as we lose contact with them. But studies suggest that when we're deprived of it, we miss nature and take steps to restore it."

Whether we listen to biologists describe the feedback between the cell and its milieu or psychologists discuss the fit between certain people and places, their words resonate with what ecologists say about the way the whole natural world works. At any level we can think of, ignoring our relationship with our environment puts both of us in jeopardy. During a century of looking inward and living in the world between our ears, we have forgotten what our forebears knew and our scientists are rediscovering. The evidence of that lapse is all around us, not just in the dirty oceans and the chemical haze that obscures the Grand Canyon but in the rising incidence of stress-related disorders. It is time to look outside ourselves and start living in the real world again.

Environmental activists urge us to "think globally, act locally," but what scientists are learning about the relationship between places and behavior suggests that thinking locally isn't a bad idea, either. If we could learn to approach larger increments of territory as we do our homes, our lives would be in better shape, and so would the planet. Because we know our psychological and physical well-being is at stake, we make sure our homes are happy, healthful, beautiful places. We pay attention to their climate and atmosphere, and are wary of incorporating technology that may endanger us and our families. We invite into them that "something in the air," often borrowed from nature, that makes places sacred. In the end, we become "addicted" to our homes not just because of their physical features but because they support our social bonds, buffer us from commotion, and help us find meaning and express our identity. Whether they are in the city, the country, or, increasingly, some-where in between, each of these sophisticated versions of the mam-malian nest is a small piece of our global village. Once we realize that just about anything that is true of our relationship with our homes is true concerning our neighborhoods, regions, and nations, then thinking locally will mean acting globally, and that means saving the world.

■

SUGGESTED

READINGS

These books and articles, most of them written by scientists whose work has been discussed in the previous pages, may be of interest to readers who wish to pursue certain topics in more detail:

INTRODUCTION

Bell, I. R. "Allergens, Physical Irritants, Depression, and Shyness." *Journal of Applied Developmental Psychology*, 1991.

Hippocrates. *Works of Hippocrates*. W. H. S. Jones and E. T. Withington, eds. Cambridge, MA: Harvard University Press, 1923.

Jasnoski, M. B. "Architectural and Interior Features Affect Mood and Cognitive Performance." *Journal of Applied Developmental Psychology*, 1991.

Lovelock, James. Introduction to James Swan, *Sacred Places*. Santa Fe, NM: Bear & Co., 1990.

Rossbach, Sarah. *Feng Shui*. New York: E. P. Dutton, 1983.

Stokols, Daniel. "Establishing and Maintaining Healthy Environments." *American Psychologist*, 1991.

Wehr, Thomas A. "Seasonal Affective Disorders: A Historical Overview." In Norman Rosenthal, ed., *Seasonal Affective Disorders and Phototherapy*. New York: Guilford Press, 1989.

CHAPTERS 1 AND 2

Booker, J., and C. Hellekson. "Prevalence of Seasonal Affective Disorder in Alaska." *American Journal of Psychiatry,* 1992.

Cook, F. A. "Medical Observations Among the Esquimaux." *New York Journal of Gynaecology and Obstetrics,* 1894.

Jamison, K. R. "Mood Disorders and Seasonal Patterns in Top British Writers and Artists." *Psychiatry,* 1989.

Kripke, D. F., D. J. Mullaney, et al. "Phototherapy of Non-Seasonal Depression." In C. Shagass, R. C. Josiassen, et al., eds., *Biological Psychiatry.* New York: Elsevier, 1985.

Lewy, A. J., H. A. Kern, N. E. Rosenthal, and T. A. Wehr. "Bright Artificial Light Treatment of a Manic-Depressive Patient with Seasonal Mood Cycle." *American Journal of Psychiatry,* 1982.

Lewy, A. J., R. L. Sack, and L. S. Miller. "Antidepressant and Circadian Phase-Shifting Effects of Light." *Science,* 1987.

Parry, B. L., S. L. Berga, et al. "Morning vs. Evening Light Treatment of Patients with Premenstrual Depression." *Abstracts of the American College of Neuropsychopharmacology Meeting,* 1987.

Rosen, L. N., S. D. Targum, M. A. Terman, M. J. Bryant, H. Hoffman, S. F. Kasper, J. R. Hamovit, J. P. Docherty, B. Welch, and N. E. Rosenthal. "Prevalence of Seasonal Affective Disorder at Four Latitudes." *Psychiatry Research,* 1990.

Rosenthal, Norman. *Seasons of the Mind.* New York: Bantam, 1989.

Rotton, J., et al. "Much Ado About the Full Moon." *Psychological Bulletin,* 1985.

Terman, M. A. "On the Question of Mechanism in Phototherapy." *Journal of Biological Rhythms,* 1988.

Wehr, Thomas, and Norman Rosenthal. "Seasonality and Affective Illness." *American Journal of Psychiatry,* 1989.

CHAPTER 3

Allport, F. H., and T. Pettigrew. "Cultural Influence on the Perception of Movement." *Journal of Abnormal and Social Psychology,* 1957.

Bell, P. A., and T. C. Greene. "Thermal Stress: Physiological Comfort, Performance, and Social Effects of Hot and Cold Environments." In G. W. Evans, ed., *Environmental Stress.* New York: Cambridge University Press, 1982.

Bell, P. A., and M. E. Fusco. "Linear and Curvilinear Relationships Between Temperature, Affect, and Violence." *Journal of Applied Social Psychology,* 1986.

Gal, Reuven, and A. D. Mangelsdorff. *Handbook of Military Psychology.* New York: Wiley, 1991.

Moran, Emilio. *Human Adaptability.* Boulder, CO: Westview Press, 1982.

Rosen, L. N., S. D. Targum, M. A. Terman, M. J. Bryant, H. Hoffman,

S. F. Kasper, J. R. Hamovit, J. P. Docherty, B. Welch, and N. E. Rosenthal. "Prevalence of Seasonal Affective Disorder at Four Latitudes." *Psychiatry Research*, 1990.

Severino, S. "High Nocturnal Body Temperature in Premenstrual Syndrome and Late Luteal Phase Dysphoric Disorder." *American Journal of Psychiatry*, 1991.

Sommers, P., and R. Moos. "The Weather and Human Behavior." In R. H. Moos, *The Human Context: Environmental Determinants of Behavior.* New York: Wiley, 1976.

Wehr, T. A., and E. Leibenluft. "Is Sleep Deprivation Useful in the Treatment of Depression?" *American Journal of Psychiatry*, 1992.

Wehr, T. A., D. A. Sack, and N. E. Rosenthal. "Seasonal Affective Disorder with Summer Depression and Winter Hypomania." *American Journal of Psychiatry*, 1987.

CHAPTER 4

Friscancho, A. R. *Human Adaptation.* St. Louis: C. V. Mosby, 1979.

Hackett, P. H., R. C. Roach, J. R. Sutton. "Medical Problems of High Altitude." In P. S. Auerbach and E. Geehr, eds., *Management of Wilderness and Environmental Emergencies.* St. Louis: C. V. Mosby, 1988.

Moran, Emilio. *Human Adaptability.* Boulder, CO: Westview Press, 1982.

Ryn, Z. "Psychopathology in Mountaineering: Mental Disturbances Under High-Altitude Stress." *International Journal of Sports Medicine*, 1988.

Suedfeld, P. "Extreme and Unusual Environments." In D. Stokols and I. Altman, eds., *Handbook of Environmental Psychology.* New York: Wiley, 1987.

West, J. B. "Human Physiology at Extreme Altitudes at Mount Everest." *Science*, 1984.

CHAPTER 5

Baron, R. A., G. W. Russell, et al. "Negative Ions and Behavior." *Journal of Personality and Social Psychology*, 1985.

Becker, R. O., and Gary Selden. *The Body Electric.* New York: William Morrow, 1985.

Friedman, H., R. O. Becker, and C. H. Bachman. "Geomagnetic Parameters and Psychiatric Hospital Admissions." *Nature*, 1963.

Ohtsuki, Y., and H. Ofuruton. "Ball Lightning." *Nature*, 1991.

Rajaram, M., and S. Mitra. "Correlation Between Convulsive Seizure and Geomagnetic Activity." *Neuroscience Letters*, 1981.

Rocard, Y. "Actions of a Very Weak Magnetic Gradient: The Reflex of the Dowser." In M. F. Barnothy, ed., *Biological Effects of Magnetic Fields.* New York: Plenum Press, 1964.

Savitz, D. A., et al. "Case-Control Study of Childhood Cancer and Exposure to 60-Hz Magnetic Fields." *American Journal of Epidemiology*, 1988.

Semm, P., T. Schneider, and L. Vollrath. "Effects of an Earth-Strength Magnetic Field on Electrical Activity of Pineal Cells." *Nature,* 1980.

Weisbrud, S. "Whales and Dolphins Use Magnetic 'Roads.' " *Science News,* 1984.

Wever, R. "ELF Effects on Human Circadian Rhythms." In M. A. Persinger, ed., *ELF and VLF Electromagnetic Field Effects.* New York: Plenum Press, 1974.

CHAPTER 6

Derr, J., and M. A. Persinger. "Geophysical Variables and Behavior: Liv. Zeitoun (Egypt) Apparitions of the Virgin Mary as Tectonic Strain-Induced Luminosities." *Perceptual and Motor Skills,* 1989.

Persinger, M. A. "Geophysical Variables and Behavior: The Tectonogenic Strain Continuum of Unusual Events (Haunts, Poltergeists)." *Perceptual and Motor Skills,* 1985.

———. "Geophysical Variables and Human Behavior: UFO Reports as Predictable but Hidden Events within Pre-1947 Central USA." *Perceptual and Motor Skills,* 1983.

———. "Increased Geomagnetic Activity and the Occurrence of Bereavement Hallucinations." *Neuroscience Letters,* 1988.

Persinger, M. A., and J. Derr. "Luminous Phenomena and Earthquakes in Southern Washington." *Experientia 42.* Birkhauser Verlag, CH-4010 Basel, Switzerland, 1986.

Persinger, M. A., and G. F. Lafreniere. *Space-Time Transients and Unusual Events.* Nelson Hall, 1977.

CHAPTERS 7 AND 8

Bowlby, J. *Attachment and Loss.* London: Hogarth Press, 1969–77.

Harlow, H. F. "The Nature of Love." *American Psychologist,* 1958.

Hofer, M. A. "Early Social Relationships: A Psychobiologist's View." *Child Development,* 1987.

———. "Relationships as Regulators: A Psychobiologic Perspective on Bereavement." *Psychosomatic Medicine,* 1984.

———. *The Roots of Human Behavior.* New York: W. H. Freeman, 1981.

Insel, T. R. "Long-Term Consequences of Stress During Development." In B. J. Carroll, ed., *The Brain and Psychopathology.* New York: Raven Press, 1991.

———. "Prenatal Stress Has Long-Term Effects on Brain Opiate Receptors." *Brain Research,* 1990.

Korner, A. F. "The Use of Waterbeds in the Care of Preterm Infants." *Journal of Perinatology,* 1984.

McClintock, M. K. "Menstrual Synchrony and Suppression." *Nature,* 1971.

Malson, L., ed. *Wolf Children and the Problem of Human Nature.* New York: Monthly Review Press, 1972.

Stern, D. "Mother and Infant Play." In M. Lewis and A. Rosenblum, eds., *The Effects of the Infant on its Caregiver*. New York: Wiley, 1976.

Thoman, E. B. "Premature Infants Seek Rhythmic Stimulation and the Experience Facilitates Neurobehavioral Development." *Journal of Developmental and Behavioral Pediatrics*, 1990.

CHAPTER 9

Barker, R. G. *Ecological Psychology*. Stanford, CA: Stanford University Press, 1968.

Putnam, F. W. *Diagnosis and Treatment of Multiple Personality Disorder*. New York: Guilford Press, 1989.

Siegel, S. "Feedforward Processes in Drug Tolerance." In R. G. Lister and H. J. Weingartner, eds., *Perspectives in Cognitive Neuroscience*, 1991.

———. "Heroin 'Overdose' Death: The Contribution of Drug-Associated Environmental Cues." *Science*, 1982.

———. "Pharmacological Conditioning and Drug Effects." In A. J. Goudie and M. W. Emmett-Oglesby, eds., *Psychoactive Drugs: Tolerance and Sensitization*. Clifton, N.J.: Humana Press, 1989.

CHAPTER 10

Cohen, S., G. W. Evans, D. Stokols, and D. S. Krantz. *Behavior, Health and Environmental Stress*, New York: Plenum, 1986.

Cohen, S., D. C. Glass, and J. E. Singer. "Apartment Noise, Auditory Discrimination, and Reading Ability in Children." *Journal of Experimental Social Psychology*, 1973.

Jones, D. M., and A. J. Chapman. *Noise and Society*. New York: Wiley, 1984.

Geen, R. G., and E. J. McGown. "Effects of Noise and Attack on Aggression and Physiological Arousal." *Motivation and Emotion*, 1984.

Heft, H. "Background and Focal Environmental Conditions of the Home and Attention in Young Children." *Journal of Applied Social Psychology*, 1979.

Lloyd, E. L. "Hallucinations and Misinterpretations in Hypothermia and Cold Stress." In B. Harvald and H. Hansen, eds., *Circumpolar 81: Proceedings of the International Symposium on Circumpolar Health*, 1981.

Rivlin, L. G., and M. Rothenberg. "The Use of Space in Open Classrooms." In W. H. Proshansky et al., eds., *Environmental Psychology*. New York: Holt, Rinehart and Winston, 1976.

Rossbach, Sarah. *Interior Design with Feng Shui*. New York: E. P. Dutton, 1987.

Seligman, M. E. *Helplessness*. New York: W. H. Freeman, 1975.

Suedfeld, P. *Restricted Environmental Stimulation*. New York: Wiley, 1980.

———. "Stressful Levels of Environmental Stimulation." In I. G. Sarason and C. D. Spielberger, eds., *Stress and Anxiety*. Halsted, 1979.

Suedfeld, P., and J. Mocellin. "The 'Sensed Presence' in Unusual Environments." *Environment and Behavior*, 1987.

Sundstrom, E. *Work Places*. New York: Cambridge University Press, 1986.

Stokols, D., and R. W. Novaco. "Transportation and Well-Being." In I. Altman et al., eds., *Transportation and Behavior*. New York: Plenum Press, 1981.

CHAPTER 11

Csikszentmihalyi, Mihaly. *Flow*. New York: HarperCollins, 1990.

Farley, F. "The Big T in Personality." *Psychology Today*, 1986.

Kagan, J. *The Nature of the Child*. New York: Basic Books, 1984.

Little, B. R. "Personality and the Environment." In *Handbook of Environmental Psychology*. New York: Wiley, 1987.

Maccoby, E., and C. Jacklin. *The Psychology of Sex*. Stanford: Stanford University Press, 1974.

Slovic, P. "Perception of Risk." *Science*, 1987.

Suomi, S. J. "Early Stress and Adult Emotional Reactivity in Rhesus Monkeys." In *The Childhood Environment and Adult Disease*. New York: Wiley, 1991.

Zimring, C. "Design for Special Populations." In *Handbook of Environmental Psychology*. New York: Wiley, 1987.

Zuckerman, M., M. Buchsbaum, and D. Murphy. "Sensation Seeking and Its Biological Correlates." *Psychological Bulletin*, 1980.

CHAPTERS 12 AND 13

Brown, B. E., and I. Altman. "Territoriality, Defensible Space, and Residential Burglary." *Journal of Environmental Psychology*, 1983.

Calhoun, J. B. "Plight of the Ik and Kaiadilt Is Seen as a Chilling Possible End for Man." *Smithsonian Magazine*, 1972.

———. "Population Density and Social Pathology." *Scientific American*, 1962.

Covington, J., and R. B. Taylor. "Fear of Crime in Urban Residential Neighborhoods." *The Sociological Quarterly*, 1991.

Fisher, J. D., and D. Byrne. "Too Close for Comfort: Sex Differences in Response to Invasions of Personal Space." *Journal of Personality and Social Psychology*, 1975.

Haney, W. G., and E. S. Knowles. "Perception of Neighborhoods by City and Suburban Residents." *Human Ecology*, 1978.

Jacobs, J. *The Death and Life of Great American Cities*. New York: Random House, 1961.

MacLean, P. D. "Brain Evolution Relating to Family, Play, and the Separation Call." *Archives of General Psychiatry*, 1985.

Stokols, D. "A Typology of Crowding Experiences." In A. Baum and Y. Epstein, eds., *Human Response to Crowding*. Hillsdale, NJ: Erlbaum, 1978.

Taylor, R. B. *Human Territorial Functioning.* New York: Cambridge University Press, 1988.

Wilson, J. J. *The Truly Disadvantaged.* Chicago: University of Chicago Press, 1987.

CHAPTERS 14 AND 15

Altman, I., and J. F. Wohlwill, eds. *Behavior and the Natural Environment.* New York: Plenum Press, 1983.

Hardin, G. "The Tragedy of the Commons." *Science,* 1968.

Kaplan, R., and S. Kaplan. *The Experience of Nature.* New York: Cambridge University Press, 1989.

Kluckhohn, F. R., and H. A. Murray, eds. *Personality in Nature, Society, and Culture.* New York: Alfred A. Knopf, 1953.

Nash, R. *Wilderness and the American Mind.* New Haven, CT: Yale University Press, 1982.

Orians, G. H. "An Ecological and Evolutionary Approach to Landscape Aesthetics." In Penning-Rousell and D. Lowenthal, eds., *Landscape Meanings and Values.* London: Allen & Unwin, 1986.

———. "Habitat Selection." In *Evolution of Human Social Behavior.* New York: Elsevier, 1980.

Sears, J. F. *Sacred Places.* New York: Oxford University Press, 1989.

Stern, Paul C., and E. Aronson. *Energy Use: The Human Dimension.* New York: W. H. Freeman, 1984.

Swan, J. A. *Nature as Teacher and Healer.* New York: Villard, 1992.

Ulrich, R. S. "View Through a Window May Influence Recovery from Surgery." *Science,* 1984.

Zube, E. H. "Perception of Landscape and Land Use." In I. Altman and J. F. Wohlwill, eds., *Human Behavior and Environment.* New York: Plenum, 1976.

INDEX